Prices, Profit, and Production

How Much Is Enough?

J. William Leasure
and
Marjorie Shepherd Turner

UNIVERSITY OF NEW MEXICO PRESS
Albuquerque

©The University of New Mexico Press, 1974. All rights reserved.
Manufactured in the United States of America by the
University of New Mexico Printing Plant.
Library of Congress Catalog Card No. 73-92995
International Standard Book Number 0-8263-0320-X
First edition

Prices, Profit, and Production

To D. S. B. and J. P. L.

Acknowledgments

While all errors, omissions, and polemical statements are certainly the fault of the authors, we would like to thank our spouses, Rosella Leasure and Merle Turner, and the following colleagues for certain valuable comments: George Babilot, Robert Barckley, Arthur Brodshatzer, David Brookshire, Charles Dicken, Arthur Kartman, M. C. Madhavan, Robert Magness, Evalyn Segal, and Douglas Stewart, as well as Reynold Hillenbrand for his stimulating ideas over the years.

We also wish to thank Elizabeth Wolkonsky and Dana Stumpf for their preparation of the manuscript—typing as well as editing.

Contents

PREFACE   xiii

1. *ECONOMICS, AMERICAN STYLE*   1
   Full Employment vs. Inflation: Facts and Ideology   4
   The Interventionists   10
   The Noninterventionists   11
   The Structuralists   13

2. *THE PROFIT STABILIZATION PRINCIPLE*   16
   Profit Constraint in Perspective   26
   The Permanence of a Large Governmental Role in
      the American Economy   28

3. *COSTS AND BENEFITS OF THE PROFIT STABILIZATION PRINCIPLE*   31
   Costs   31
   Benefits   40
      Fiscal Policy   41
      Monetary Policy   43
      Long-Range Objectives   45
         Individuals   45
         Businesses   45
         Government   46
   Role of PSP   48
   Freedom and Control   48

4.	*ECONOMIC EVOLUTION AND THE PUBLIC INTEREST*	51
	Level of Performance	52
	Conflict of Goals	55
	More Goods	57
	Different Goods	60
	More Jobs and Different Jobs	62
	Principle by Which To Judge Performance: Public Interest	64
5.	*STABILITY WITHOUT GROWTH IN THE EIGHTIES?*	67
	Declining Growth of GNP	75
	Conclusion	78

APPENDIX OF RELATED READINGS   81

Chapter 1

  Reading 1. Excerpts from a Debate between Milton Friedman and Robert Solow upon the Occasion of the Johnson Wage-Price Guideposts   82

  Reading 2. The Issue of Whether Monopoly Is an Economic Problem and the Division among Economists on What To Do about the Current Structure of American Industry   85

Chapter 2

  Reading 3. Issues in Freedom and Control: What Economists Think about Planning and Public Interest   91

  Reading 4. The Problem of Inflation as Arthur Burns Sees It   95

  Reading 5. Shortages, PSP, and Price Control   100

Chapter 3

  Reading 6. Excerpts from the British White Paper of March 1973, "The Counter-Inflation Programme," "The Operation of Stage Two," Presented to Parliament by the Chancellor of the Exchequer by Command of Her Majesty, March 1973   103

  Reading 7. Social Responsibility of Corporations   105

Chapter 4
   Reading 8. Guaranteed Income Plans and
      Some Alternatives   109
   Reading 9. Issues in the Solution of the Problem
      of Unemployment: The Federal Government's Role
      as Employer of Last Resort   117

Chapter 5
   Reading 10. Optimal Population (A Spaceship Approach) 124

NOTES   127

BIBLIOGRAPHY   139

INDEX   141

Preface

Why This Book Was Written

The study and teaching of economics have convinced us that our science has more to offer than has been realized by most people. While professional economists are sought as advisers to business and Congress and as members of the president's Council of Economic Advisers, their advice tends to be ignored by those same congressmen, presidents, and businessmen who persist in thinking that economists are impractical people.

Even so, since World War II economists have had increasing public exposure. Congressional committee reports may involve thousands of pages and many hundreds of economists, but often what they write and talk about is simply a reissue of an old journal article or book, regardless of the subject the congressional committee is investigating. Economists crop up more and more in popular periodicals, but with limited effectiveness, as they address themselves only to the "economics" of the problem, whether it is poverty, price stability, unemployment, or balance of payments. Few make any effort to relate the "professional" problem to its political, social, and psychological aspects. The economist is generally an intelligent informed person and should feel free to speak to the social and political aspects of societal ills. Indeed, unless he is willing to do so, no matter how economically sound his proffered solutions, he will still be regarded by the public as an *inefficient* problem solver.

It is more the attitude of the public than the inadequacy of the science of economics that has kept economists in the ivory tower, chattering to each other about esoteric topics. And many members of the American economic profession do descend for brief efforts at public persuasion, the most prominent being Paul Samuelson, John K. Galbraith, and Milton Friedman. Unfortunately, what generally occurs is that their differences on matters understandable to the public are aired, and the "practical" people then impatiently dismiss these issues without trying to understand what is behind all the differences. After all, if economists cannot agree among themselves, who needs to listen?

This book, like those of others, is an attempt to persuade. But it is our hope that we also provide the tools to help the student and general reader understand the technical arguments that economists are having among themselves. Essentially our argument is a positive one:

1. That the economy is not functioning as well as it might and therefore as it ought to do. What is meant by "not functioning well" is the inability to achieve full employment with reasonable price stability.
2. That this fact is generally known to economists and that they all agree that "something must be done" without agreeing on the "something."

   We add our own proposal of the Profit Stabilization Principle to those of others and try to show its costs and benefits and how it might work concurrently with some of the other proposals. This proposal makes price increases contingent on profit rates.
3. That the short-term performance of the economy must be seen in the context of various and alternative performance standards, not limited to full employment and price stability. We try to explain in a historical, evolutionary context such standards as engineering efficiency and cost efficiency, as well as the performance we think people want today from their economy.
4. That a judgment principle must be brought to bear on the choice among the alternative standards of performance. We take the appropriate one to be the public interest in the best possible and most equitable performance within a climate of maximum freedom and minimum controls on the individual. This would include a more equitable distribution of income along with full employment and price stability.
5. That the long-run performance of the economy, even the future which our children and grandchildren face, requires

that we take into account certain developing problems, such as the pressure of population and the implications of the kind of unplanned economic growth which is ordinarily associated with full employment. We argue that programs like our Profit Stabilization Principle will not only help solve short-run problems but will make a contribution toward facing up to the long-run problems as well.

Thus our book, while attempting to persuade, is also directed toward greater understanding of the central performance standards that might be applied to the economy. We want to stimulate thinking and debate rather than offer a panacea. But if the reader will forgive us for moralizing at times, he will understand our central theme, which is that the American economy has come as far as it can on unbridled greed. For the sake of all of us, capitalist and worker alike, the American economy needs some fundamental reordering of economic motivation. What are we waiting for—a revolution?

1
Economics, American Style

What is wrong with the United States economy? If you asked twenty people, you would probably get twenty answers ranging from "nothing" to some expression of a need for a revolution. But if you asked twenty economists, the answers might not be quite so disparate. Chances are that they would find some common ground on these propositions: (1) while the economy is fundamentally sound as an operating machine, its performance might be improved in the direction of making it more equitable; (2) the most immediate and promising way to do this is through provision of full employment without inflation; (3) an appropriate pace of economic growth would make a sizeable contribution toward these ends.

This is what you might identify as the party line or orthodox position of the American economist. Were there a radical economist among the group, he would probably be dissatisfied with all these propositions. He would deny the claim of fundamental soundness. He would consider the possibility of "improving" the economy by promising full employment without inflation another opiate of the people. His position would be one that fundamentally challenges orthodox economics. Ordinarily, such a person would not propose reforms, as he would be either a cynic or a revolutionary.

Within the group of orthodox economists, there are also differences which are traceable to different value structures. For example, while most economists, when pressed, will admit some inequity in the distribution of wealth, very few would care even to discuss its redistribution. This has been the accepted position ever since John Bates Clark of Columbia made his point that the distribution of *wealth* must be taken as a given before the economists could scientifically discuss the distribution of *income*.[1]

As for the distribution of income, most orthodox economists

would also be ready to accept it as a given, not to be altered. They would point out that no society, including a socialist society, approaches anything like equality in the distribution of income. Sweden and Great Britain have income distributions similar to ours.

The distribution of goods and services would, however, bother some of the orthodox economists. Realizing that the distribution of goods follows the distribution of wealth and income, they would be faced with a dilemma. Wealth and income monopolize opportunity in our society. Sometimes education and personal skills may make it possible for the individual to break out of a ghetto into the mainstream of American life. Usually, however, education of good quality is one of the opportunities that has been reserved to those possessing wealth and income. That, of course, is the economic nub of *Brown* vs. *Board of Education* and other school integration suits.

The way out of this dilemma for the typical economist is usually the adoption of that third proposition: that an appropriate pace of economic growth will provide new opportunities and more income. "The pie will get bigger" is the argument, and so the number of goods available to all members of the society will be increased. Yet this acceptance of the program of economic growth brings its own dilemmas. These may be seen as long-run problems (see chapter 5), which are real and frightening, and as the short-run problem of trying to achieve full employment without inflation. Thus, while orthodox economists may agree on the desirability of achieving sufficient growth to provide full employment on the one hand, they also see it as a threat to price stability. They cannot easily agree on whether or how these goals may be achieved simultaneously. Their differences are traceable to three overlapping areas—economic theory, ideology, and the facts.

Historically, since Adam Smith, the mainstream of economic analysis has taken the position that the economy would operate according to some natural laws provided there was no outside interference. If each person were left to pursue his own welfare with the least interference from government, then that would produce the best of all possible worlds. Competition existed to bring this about, and social theories like Darwinism made the late-nineteenth-century economists even more certain than Adam Smith of a happy outcome from laissez-faire. Smith had serious reservations about some forms of business power, such as corporations.

As theory developed, it appeared to most economists that marginal decisions by individuals and firms in competitive markets provided this harmonious world. Economists who disagreed found it hard to get university positions. Of major nineteenth-century economists, only John Stuart Mill counseled that it was possible to affect the division of income without upsetting the productive machinery. And only in this century, with the Keynesian general theory, did orthodoxy take up the new position that laissez-faire policy *might* be inadequate to provide full employment. This was admitted only when economists were confronted with the example of the new Soviet society and the suffering generated in the West by the Great Depression.

Although theory no longer requires a laissez-faire attitude on the part of economists, orthodox American economists remain reluctant to attempt to substitute their judgment for that of the market. Thus, considerations of whether to do something about an economic problem and what to do about that problem are now matters more of values or ideology than of theory.

Even "objective" economists have biases based on their values, which in turn are generated primarily by the environment in which they live. That environment is generally rather traditional and accepting of the *status quo.* Furthermore, when it comes to the question of full employment and price stability, the facts of economic life do show a conflict between the two when the unemployment rate is 3, 4, or even 5 percent of the labor force.

This book concerns itself with a reordering of the economy so that: (1) in the short run, economic growth will not end in inflation even before we achieve full employment; (2) in the long run, growth does not lead us to the destruction of our physical environment and possibly ourselves, but to the enrichment of our human development. What we suggest will seem revolutionary to some; to others, inadequate to solve our problems. We are not proposing socialism, but we are proposing that the rules of the economic game be altered with respect to profit. We suggest that profits of most corporations, and not just those of public utilities, be brought under social control in the sense that price changes be made dependent on profit rates.

Before we discuss our proposal for the achievement of full employment with price stability, we wish to examine in greater detail the historical facts on employment and inflation in our economy.

Full Employment vs. Inflation: Facts and Ideology

The trade-off between unemployment and inflation has been the most recognized and discussed problem of the American economy since 1960. The most popular presentation of this is in terms of the Phillips curve. This curve plots the relationship between unemployment rates and prices. As one would expect, when unemployment rates fall, prices rise.

What seems to be happening in our economy is that when we approach full employment, companies realize that even if prices were higher, they could sell at least as much as they have been selling in the recent past. With a higher income level for everyone, they may even expect to sell more products at the higher price. When a few large companies dominate an industry, they are able to set the price. With profits as the primary goal, they usually raise their prices in order to raise their profits. After this phase comes the realization by the labor unions that the purchasing power of wages is declining. Even if money wages have risen, what they can buy has not increased. Recently this has come to be called "expectation inflation."[2]

In the next phase comes the push by the unions for higher wages. The companies cannot grant pay increases without a reduction in profits, but to avoid a prolonged and costly strike they may grant the wage increases provided they can also raise their prices. In this way profits remain high, wages rise, and prices continue rising. The people that are hurt are the other wage earners or pensioners whose incomes don't rise, or else fail to rise as much as prices. This circle

Economics, American Style

repeats itself unless some outside force, such as the government, intervenes. Government can act to create a recession, but even with a recession price stability may not be achieved. From 1955 to 1958, unemployment rose from 4.4 percent to 6.8 percent, yet the consumer price index rose from 93.3 to 100.7 (1957–59 = 100).

This example from the 1950s illustrates the changing relationship between prices and unemployment between the 1950s and the 1960s. In the earlier decade, there was not a close relationship, if any, between percentage change in prices and in unemployment rates. In the 1960s, however, the points do fit a curve fairly well. Figure 1 plots these points for the longer period of 1950–69 and for the shorter period of 1960–69.

When we examine the more recent years 1970 and 1971 however, we find that a high unemployment rate is now associated with a high rate of price increase. Either the relationship, if any ever did exist, is once again tenuous, like that in the 1950s, or else the 1970s will give us a new relationship between prices and unemployment. Some economists accept the latter explanation and argue that something has changed in this society, so that a new "trade off" must be postulated. Reducing unemployment to 3 percent would now require our sustaining a 5 percent or greater annual rise in prices. In the same way, more modest proposals for lowering unemployment rates to 4 percent would now require the acceptance of a better than 4 percent annual rise in prices. The new and the old trade offs as postulated are shown in figure 2.

Economists agree that many problems, such as poverty, welfare cases, racial integration of the labor force, equal opportunity, and so on, cannot be solved without full employment. Many economists feel that the cost of buying some price stability at the expense of leaving these problems unsolved is too high. One group, the unemployed, has economists like John Kenneth Galbraith to champion its demand for more jobs. The other group, the employed, has economists like Milton Friedman to present the case for price stability. The supporters of price stability are prevailing in the early seventies, and the consequences are that people have had to accept more unemployment even though this has not avoided inflation. Since the unemployed are primarily young people,

FIGURE 1

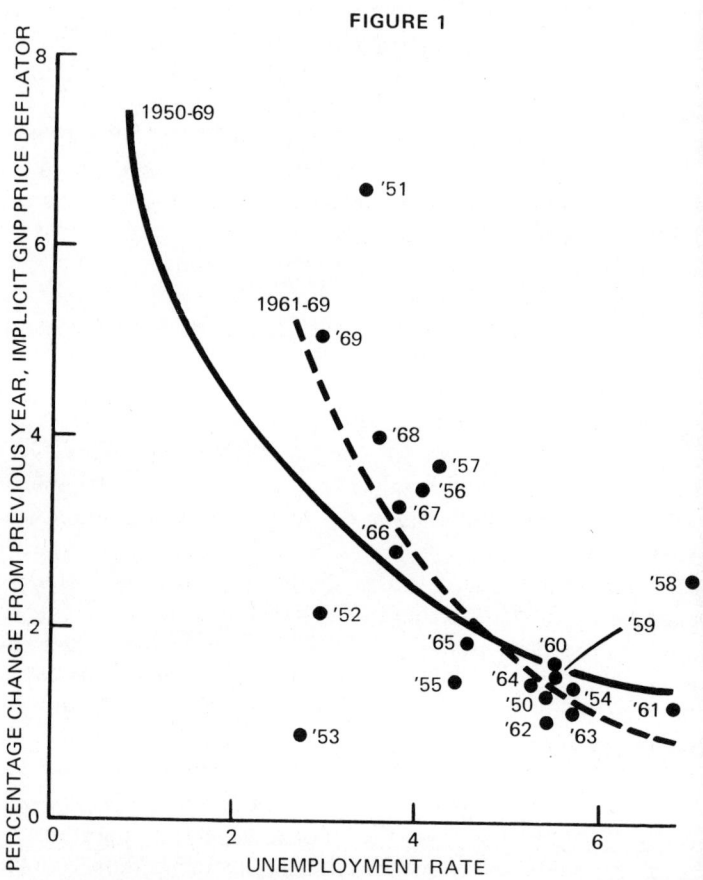

The rate of increase in prices has usually accelerated when unemployment was at a very low level. The black line represents an overall average approximation of this trade-off pattern over the entire period 1950-68. The scatter of dots showing the inflation-unemployment relationships for individual years around this longer-run pattern indicates the lack of a precise statistical fit. On the other hand, the rates of advance in prices and movements in the unemployment rate for each year between 1961-68 are represented remarkably well by the overall pattern for this period—the [dotted] line. These relationships are the implicit basis for most forecasters suggesting that some increase in unemployment must be accepted in order to reduce inflation of 1969. Source: *Monthly Review,* Federal Reserve Bank of Atlanta, February 1969, p. 21.

Economics, American Style

FIGURE 2
PHILLIPS CURVE
Trade-off between Inflation and Unemployment

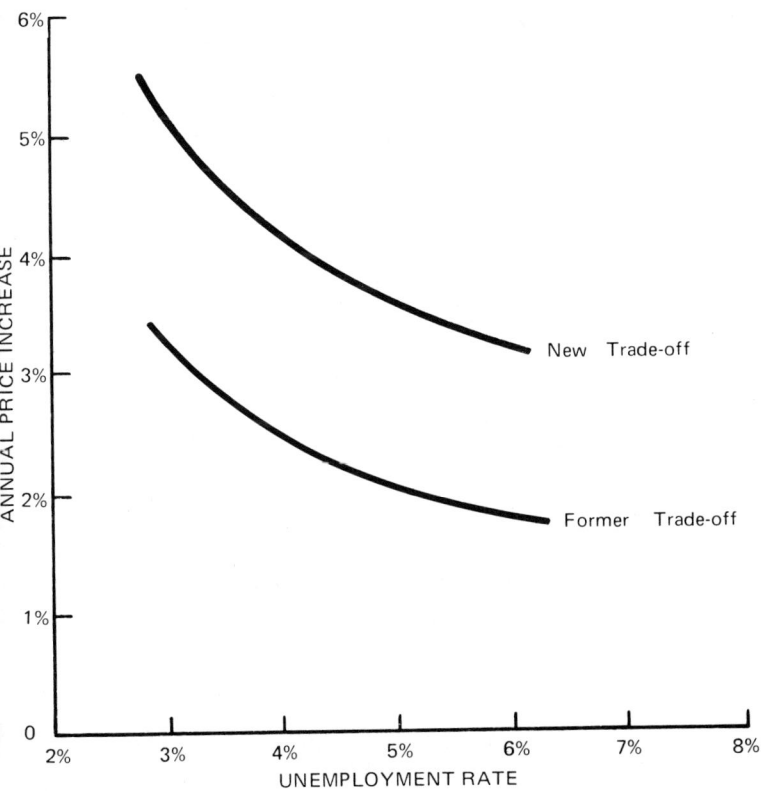

Source: George L. Perry/Brookings Institution, in *Newsweek,* January 24, 1972, p. 54.

women, and the disadvantaged, some economists ask: Why worry about a 5 percent unemployment rate, especially if anything less than 5 percent means more inflation?

This latter argument has two steps: (1) Such groups as the housewife (who already has someone to support her) and the teenager (who ought to be in school) and the structurally unemployed (which includes the unemployed aerospace engineer as well as the disadvantaged) are swelling the ranks of the unemployed. (2) Therefore, all of us who are now employed (which is, after all, most of us), can simply agree that a 3 or 4 percent unemployment rate is unrealistic, and that 5 percent is good enough after all. This is what we did during the fifties when we decided that a 3 percent rate was unattainable and switched to 4 percent as the "desirable" unemployment rate. (For comparison, the typical European rates are generally below 3 percent on a similar and perhaps superior counting system, in spite of the fact that most European young people normally enter the labor market at 15–17 years of age on a full-time basis.)

From 1950 through 1969, a period of twenty years, the United States unemployment rate fell below 4 percent in only seven of those years. This occurred in the early 1950s and the late 1960s. During each of these periods there was a war—the Korean War and the Vietnamese War. In peacetime the rate often rises to more than 5 percent. So this rate is a common one, but its occurrence has always been heralded as a recession.

In addition to the unemployed, there are people who are underemployed, that is, persons working twenty or thirty hours a week. Many of these people want full-time employment. In 1969, for example, there were 2.8 million persons unemployed and another 2 million underemployed. Even among the employed, there are the "working poor." This shows that the unemployment figures are quite likely to understate the real impact of unemployment. For example, Neil Chamberlain and Donald E. Cullen estimate that in that same year, 1969 (the last year before the 1970 recession), a total of about 11 million persons were jobless at some time or other during the year.[3] Half were unemployed for fewer than five weeks, but nearly one-fourth were out of work for fifteen or more weeks. Furthermore, the burden of this unemployment falls differentially upon the young and the unskilled, and particularly upon those other than whites (including adult males) who generally have unemployment rates which are at least double those of the general rate.

As for women, including married ones, rising participation rates are a long-term historical trend, beginning at least as early as the turn of the century, when the labor force participation rate of women fourteen years and older was 20 percent. Now women have a participation rate of 42 percent, and their share of the total labor force is about 38 percent. While there has been some acceleration of this trend since World War II, this is due in part to changes in family structure and the educational patterns of women, and more recently to the change in birth rates, all of which are interrelated.

The argument regarding teenagers' unfairly swelling the ranks of the unemployed has even less merit. The long-term trend is that the young person enters the labor market later than his father or mother had done. Thus, the labor force participation rate of the young is lower than in 1947. For example, labor force participation rates for males eighteen and nineteen years old dropped from 80.5 percent in 1947 to 69.9 percent in 1970.

The proposal that we "solve" the problem by once again changing our level of aspirations so that we no longer strive for an unemployment rate of 3 to 4 percent is obviously not a "solution." We cannot accept a 5 percent unemployment rate as normal when such problems as integration and welfare caseloads keep nagging us.

And the authors do not accept a 4 percent rate as normal. Since economists differ in their opinions, and sometimes even on facts, we conclude that the crux of this matter is partly ideological. Our position is that we must achieve the traditionally accepted goal of an unemployment rate of 3 percent. This definition of full employment takes into account the fact that there will always be some people who are temporarily unemployed while seeking a better job or moving to another location. Our proposal is concerned with achieving this goal. Other economists, however, have different proposals. We should, therefore, examine their recommendations first and evaluate them in terms of this goal. One recognizable division among economists, partly based on ideology, is between the interventionists and the noninterventionists.

The Interventionists

John K. Galbraith represents the interventionists, in that he has for some time counseled the government to expand its role in price setting. Galbraith, however, would not have the government try to set all prices, but only in those industries in which there is concentration of power or on those commodities which are truly basic to the standard of living of the American public, such as food and clothing. This would, he believes, take away the justification for cost-push inflation where workers demand higher wages to meet the rising cost of living.

It was a great surprise not only to the nation, but also, apparently, to the president's chief economic adviser, Paul McCracken, that President Nixon also turned up in the interventionist camp. McCracken had written an article in July 1971 outlining why the Nixon administration would not intervene directly in price control, only to have his boss announce a different policy in August 1971. Paul Samuelson and other economists soon made public statements of support.

While there was agreement among these spokesmen that some direct controls were desirable, Galbraith and Nixon disagreed on what was to be controlled. Galbraith meant to emphasize control on the prices of what we eat and wear, and of our transportation to work. He would even go as far as a subsidy to keep prices of these things within the reach of working people. His understanding of what might be done is enhanced by his experience in price stabilization during World War II. At that time, it was necessary for the public to subsidize the production of milk and butter, for example, in order to keep the prices of these things low. On the other hand, these low prices helped to make wage stabilization acceptable to the working population. Galbraith felt it was a good lesson.

President Nixon in Phase I went about it another way. At first, prices and wages in every industry were to be controlled. But then it was learned that *some* prices (gasoline, seasonal fruits and vegetables, meats, and so on) were *not* to be controlled, and even some services were exempt—such as brokers' fees. There were, of course, very good reasons offered for this pattern. But many of the working people, especially those in unions whose contracts had not been negotiated before the freeze, were caught in a squeeze of rising food prices and semifrozen wages. Hence, phases II, III, and IV became increasingly complex, if not confused.[4]

These kinds of problems are foreseeable and account for the fact that most economists at most times are reluctant to begin intervening in prices. In this sense, Galbraith was out of step with most economists in urging intervention even during Johnson's administration. He remained out of step by supporting his own brand of price controls when Nixon's interventionism came along.

The major problem associated with price and wage controls is that each case must be examined individually. Each company and each union, therefore, can and does exert tremendous political pressure to achieve its goals. Furthermore, if price controls are to be effective, a large bureaucracy is necessary to gather and evaluate all the data pertaining to each industry. A tremendous amount of knowledge and understanding is required for equitable decisions.

The vulnerability to political pressure of a price and wage board makes it a relatively unsatisfactory substitute for collective bargaining and the influence of the market on wages and prices. This is partly the reason most economists are opposed to long-term use of such controls. It is only natural that the inability to apply objective criteria in each case would lead to political favoritism. As will be seen in chapter 2, our proposal does not regulate prices or wages, but makes price increases dependent on profit rates and lets the company and union bargain over wage increases.

The Noninterventionists

Milton Friedman champions the noninterventionists. He indicates that the economic system will achieve a reasonable performance level on all counts, including the number of jobs and price stability, if and only if his prescription for the reliance on monetary controls is adopted. The real cure-all, in his view, is the steady and controlled rate of increase in the quantity of money permitted by the Federal Reserve banks. Everything else would then have to adjust to this rate, which itself is related to the increase in population and productivity. Workers might demand more wages, but since the increase in the supply of money would be fixed, all prices could not rise simultaneously and still permit the purchase of the goods. Since employers, consequently, could not successfully pass on wage increases in the form of price increases, excessive wage demands could not be met. Prices would remain stable.

Friedman argues simply that inflation is the result of too much money being created by the banking system with the acquiescence of the Federal Reserve system under pressure from the business community, the government, and the consumers for the financing of their deficits. This argument forms the basis for his belief that interventionism in the form of government controls is merely a futile gesture which will further distort the allocation of resources. To follow Friedman's advice, however, would mean the acceptance of high unemployment rates. Given the fluctuations in our economy resulting from wars, international recessions, the clustering of inventions and innovations, and inadequate demand, most economists believe that the Federal Reserve must respond to such changes in our economy by permitting changing rates of growth in the money supply. If the Federal Reserve did not do so, then a slowing down of the economy for one reason or another would merely continue unemployment for a longer period of time than would be the case with a more active monetary policy. Now when we have a recession, the Federal Reserve makes an extra effort to stimulate the economy via a rapid expansion of the money supply. Under Friedman's plan, it would be business as usual regardless of the state of the economy.

In summary, then, orthodox economists' prescriptions for the disease of inflation accompanying rising employment levels range from control of prices and wages (but not profits) to control of the supply of money. The prescription, however, depends on the disease. Another group, the structuralists, have a different explanation as to why we now seem to need a higher level of unemployment in order to maintain some semblance of price stability.

The Structuralists

The structuralists are the people who decry the present organization or structure of business: more specifically, the concentration of power in the hands of a few giant corporations. Economists like Walter Adams, who have concerned themselves with problems of the real rather than the theoretical working of the economy, claim that some major changes in the pricing of goods in our economy can no longer be ignored. For Adams, the competitive conditions are rapidly disappearing. Chapter 3 will explain in depth the implication of this for performance standards.

Adams's observations of industries such as steel tell him that the present structure of American industry leaves the producer free to charge what the traffic will bear.[5] Such prices cannot reasonably be expected to have any given relationship to costs regardless of what the industry argues. Therefore, what may be behind the problem is a concentration of economic power so excessive that the rate of unemployment necessary for stable prices will continue to rise along with the growing concentration. Unless we are willing to curb such power, then the mere acceptance of 5 percent instead of 4 percent as the "desired" unemployment rate is not likely to work better than our substitution of 4 percent as a goal for 3 percent.

The domestic system of producer control of prices is reinforced by internationalized private cartel agreements between firms. These agreements effectively divide the American as well as the world market among the major Western producers. While governmental tariffs and quotas are economically puny beside such a system of private controls, they do help, nevertheless, to reinforce the international cartel system. And industries like steel which have been unsuccessful in establishing private controls generally seek governmental protection from foreign competition. Others, such as the oil industry, mesh an elaborate system of public controls on production with federal government import quotas. In such a system, one may ask if there is any level of unemployment which might produce prices which are other than "what the traffic will bear."

More recently, another part of the structuralist argument has been taking form. Industry as it is now organized not only sets prices, but has also been shown to contribute directly to unemployment, via the conglomerate.

The conglomerate movement of the 1960s was a relatively new phase in the American monopoly pattern. In the past, industry either merged vertically, as did Henry Ford when he bought rubber plantations and attempted to make every part of the final Ford product, or horizontally, as did the chain stores and supermarkets which bought up the stores of their competitors. The new movement toward conglomeration exhibits no such obvious pattern. Its effects on the domestic economy are only now beginning to be understood, but some evidence indicates that this movement results not only in fewer firms but also in fewer employees in the remaining firms. This may be the result of economies of scale or another example of cost-cutting efficiency. In either case, however, the trend is not favorable for those needing and seeking scarce jobs.

In summary, the structuralist may now argue that the appropriate remedy to the inflation/unemployment problem is a renewal of efforts to enforce antitrust aims, which would prevent monopoly pricing and also undermine the conglomerate destruction of jobs in absentee-owned industries. Price competition from freer international trade might also help.

All of these ideas and proposals show that nearly everyone agrees that the present system is somehow intolerable. Something has to be done—something effective. Friedman's solution is not likely to be tried, simply because few economists and probably no politician would accept the poor level of performance in terms of the high unemployment rates that Friedman's scheme is likely to bring intermittently. Direct controls over some prices and wages as in Nixon's Phase II are considered a stopgap at best. The structuralists are correct in pointing out that there are fundamental reasons in our economic make-up for this inflation with unemployment bias.

In this context of debate, we make a proposal that promises many benefits with few costs, for it promises only to tame one major problem (inflation) so that the other (unemployment) may be treated according to present knowledge. Our plan, known as the Profit Stabilization Principle (PSP), need not be the only program used, however. For it is compatible with the structuralists' antitrust enforcement, as well as with Galbraith's standard-of-living controls and even with Friedman's utilization of a predetermined and automatic monetary policy.

2

The Profit Stabilization Principle

We think that full employment (3.0 percent unemployed) and price stability are the immediate goals sought by the majority of Americans at the present time. One way to achieve these goals is to enforce wage and price controls. As noted earlier, however, these are difficult to enforce and are often inequitable because the wage and price boards respond to political pressures from the strongest lobbies. Furthermore, such controls usually exclude profits from any regulation. What is preferable, in our opinion, is an explicit set of rules establishing an automatic procedure which would permit price increases only under certain conditions. We propose that profit rates should be that guide.

Profit is defined as the excess of measurable monetary return over measurable monetary costs associated with some economic activity. It can therefore be applied usefully to any form of economic organization. Profit serves many functions in every economic system. Only the name by which it is designated may differ. Each economic activity has an optimum rate of profitability consistent with its objectives. In our economic system, large profits may be consistent with full employment and growth, but in recent years, they have not been associated with price stability. In the case of private undertakings, profit is usually associated with the immediate aims of cost efficiency, a term which will be discussed at some length in chapter 4. All we need to say here is that *unlimited* profits by American corporations, particularly the large ones, may severely restrict the maintenance of the objectives of full employment and price stability.

The government has changed its role in the economy so that it now strives to maintain a system without economic depressions so that well-operated firms are virtually guaranteed a minimum profit. Yet industrial concerns are operating in these new circumstances as though the uncertainty were the same as it was during the 1920s. They use their economic power to charge what the market will bear, which may have been reasonable under the old rules where the market was such an uncertain phenomenon. Their failure to adjust to the new condition, which no longer requires such greed, simply makes it more difficult for the government to maintain favorable markets via full employment conditions because of the inflation constraint.

We wish to alter this trend by making price increases contingent upon profit rates in the following way: Only if profits fell below a specified rate could companies raise their prices. And, if profits exceeded a certain rate, they would be taxed 100 percent. This would force companies to consider price reductions when profits were high. The profit rate would be based on an average of these returns for the five previous years. Since profits would tend to fluctuate within this range, we call this proposal the Profit Stabilization Principle (PSP).

Price increases would depend on profits. Wage increases would be guided by these price and profit constraints, but labor and management would be free to bargain over wages. This is the kind of reordering of our economy that is necessary at this time. The following paragraphs give a more detailed explanation of PSP.

In industries where there is limited competition because of a relatively small number of companies—such as steel, automobiles, and chemicals—prices could be increased only when the profit rate after normal taxes (local, state, and federal) dropped to some low rate such as 6 percent. Likewise, all profits in excess of a 10 percent return after normal taxes would be paid in full to the government. The lower and upper limits of 6 and 10 percent are somewhat arbitrary and could just as well be set at 7 and 12, for the range could vary according to the size of the corporation, as will be discussed later in more detail. But this range from 6 to 10 percent is related to the long-run average rate of return in the United States after taxes, which is between 8 and 10 percent.

For example, according to the *Fortune* magazine directory in

1962, General Electric's profits (after taxes) were 15.4 percent of invested capital; U.S. Steel's were 4.9. In 1965, they were 16.9 and 7.6 respectively. In these two years, the comparable figures for Standard Oil of New Jersey were 11.1 and 11.9. The median return for the 500 largest industrial corporations in 1962 was 8.3 percent, while firms like Great Atlantic and Pacific Tea and Safeway earned 11.2 and 13.3 percent respectively.

The highly concentrated industries could be identified by some type of concentration ratio, e.g., the proportion of total sales attributed to the top four companies. For example, if four or fewer companies controlled over 40 percent of the sales in that industry, then they would be regulated in the manner described.[1]

The profit rate is simply net profits after normal taxes (local, state, and federal) divided by assets. Assets are determined at their current value and include all tangible assets such as cash balances, accounts receivable, patents, copyrights, stocks, and bonds. It is possible that different types of industries might require different definitions of assets appropriate to their activity.

Depreciation of buildings and equipment would be allowed according to one of the standard procedures now permitted by the Internal Revenue Service. The calculation of profits, therefore, would follow the guidelines already approved by the government. The estimation of the current value of all tangible assets might pose more problems. Periodic assessments of all buildings and equipment would be required, with annual adjustments or increases in value via inflation or market changes. These estimates are now made in the preparation of statements of net worth and for property tax purposes. There is some control now over these procedures, but perhaps PSP would call for an extension of them in terms of standards and supervision.

The profit rate used as the guideline for price increase does not refer to the rates for the current year only. It refers to the average return for the previous five years. In other words, a moving average is used. Each year the rate of return for the previous five years is calculated to obtain the average for tax purposes. The two tables below will illustrate this point. Table 1 shows profits for a hypothetical company after normal taxes as a percentage of assets

TABLE 1

Company A's Net Profits as Percentage of Assets

Year	Assets	Profits After Normal Taxes	Net Profit as Percentage of Assets
1971	$ 900,000	90,000	10.0
1972	925,000	64,750	7.0
1973	975,000	97,500	10.0
1974	990,000	128,700	13.0
1975	1,000,000	100,000	10.0
1976	1,000,000	110,000	11.0
1977	1,050,000	52,500	5.0
1978	1,060,000	31,800	3.0
1979	1,060,000	53,000	5.0
1980	1,100,000	55,000	5.0
1981	1,150,000	88,000	8.0
1982	1,200,000	149,500	13.0
1983	1,250,000	144,000	12.0

for individual years. Table 2 shows them for five-year periods where a moving average is used. In effect, a five-year average of profits is divided by a five-year average of assets.

In table 2 we see that company A could raise its prices at the end of 1980 and again after 1981. The five-year average for profits ending in 1980 is 5.7 percent and in 1981, 5.2 percent. Both are below the 6 percent minimum level which permits a price increase.

If we look at table 1 for data on this hypothetical company, we find that profit in 1980 generated a 5.0 percent return. Let us assume that the company would then raise its prices and achieve an 8.0 percent return in 1981. The five-year average ending with 1981, however, is 5.2 percent, which is also below the minimum. In 1982, therefore, the company can raise its prices once again. There is one further constraint, however.

The price increase has to be geared to that required to earn no more than 12 percent the following year. If the price rise was such that profits rose above 12 percent in the next year, then a price reduction would be required in the following year and thereafter until the return was no more than 12 percent. An upper limit of 12

TABLE 2

Company A's Five-Year Averages of Net Profit as Percentage of Assets[+]

Five-Year Periods	Average Profit Percentage
1971-75	10.0
1972-76	10.2*
1973-77	9.7
1974-78	8.3
1975-79	6.7
1976-80	5.7**
1977-81	5.2**
1978-82	6.8
1979-83	8.5

[+] These are based on a moving average. For example, the return for 1971-75 is:

$$\frac{\$90{,}000 + 64{,}750 + 97{,}500 + 128{,}700 + 100{,}000}{\$900{,}000 + 925{,}000 + 975{,}000 + 990{,}000 + 1{,}000{,}000}$$

The result is the same as if the numerator and denominator had each been divided by five.

* Excess profits must be paid to the government.

** Prices can be raised at the end of these periods.

percent rather than 10 percent is permitted in the following year to permit the company to offset some of the lower profits in the earlier years.

The purpose of this limitation on price increases after profits have dropped below the minimum level is to prevent excessive profits in the following year. Such an increase in profits might provoke unions to bargain for excessive wage increases. The union could argue that the higher-than-usual wage demands would merely reduce the excess profits that would ultimately be paid to the government anyway. If this practice prevailed, then we would have simply another form of cost-push inflation.

Let us assume that company A chooses to raise prices once again with the unforeseen consequence that the profit return for the year of 1982 is 13.0 as seen in table 1. This return exceeds the maximum return of 12 percent in the year in which a price increase occurs. Our scheme would consequently require that prices be reduced to such a level that the next year the return was no more than 12 percent. In our hypothetical example, such would be the case with company A, as the return in 1983 drops to 12.0 percent. If it had been higher than 12.0 percent in 1983, then company A would have to continue making price reductions until the return fell to 12 percent or less.

If the five-year profit return is pushing above 10 percent, then unions have every incentive to bargain for higher wages, and the company may as well grant some of their demands because all profits in excess of that have to be paid in full to the government. The company may also wish to consider possible price cuts. If profits are at a rate of, say, 8 or 9 percent, the unions and management will have a struggle. The reason is that at a rate of return over 6 percent, the company cannot raise its prices. Therefore, if excessive wage demands were granted, prices could not be raised given the current 8 or 9 percent return used in this example. Consequently, at some future date profits might fall to an unacceptably low level, such as 6 or 7 percent, which would still not permit a price increase. The company, therefore, has a strong and persuasive argument for resisting wage demands much larger than the gains in productivity for labor.

When profits did fall below the 6 percent level, then the company could raise prices as in the example described earlier. In this case, it is possible that unions and management might play a game whereby the company deliberately granted excessive wage demands, profits fell below 6 percent, and then prices could be raised in order to restore the average return of 10 percent. But this is not very likely, since prices could be raised only if profits fell below the minimum when averaged over a five-year period. The following example will illustrate this.

Suppose that company A's profit rates after normal taxes for each of the previous five years have been 10, 11, 5, 3, 5 percent, as seen in table 1 for the years of 1975 through 1979. The five-year average of these is 6.7 percent. Even though profits have been relatively low for the previous three years, the average rate is still 6.7 percent. Profits would have to continue for the next three years at a rate of about 5 percent in this example before the company would be permitted to raise prices. If this pattern had been the result of a strategy, a "high price" would consequently be paid in order for the company to be allowed to raise prices so as to earn the 10 percent return once again. Such a game would require too long a waiting period before the payoff. There are other deterrents to this game which will be examined in the next chapter. Because of all of these, it appears that the company and the union would restrict their collective bargaining to the range determined by the current prices, profits, and trends in productivity, with a rate of return of 10 percent as the guideline.

The inherent advantage of this proposal as opposed to direct controls is that labor and management are permitted to engage in collective bargaining for wage increases even though price and profit constraints are imposed. The price increases that do occur are the result of an objective criterion, i.e., the rate of return on assets, and not the consequence of a political decision by a price control board. All that is required is an explicit and unchanging system for the determination of profits and current value of assets akin to the rules now established by the Internal Revenue Service and assessors, as well as a board for ensuring the observance of these rules.

At present the Internal Revenue Service determines how fast buildings and equipment can be depreciated. The value of property is assessed regularly for local tax purposes. These rules and criteria could be extended somewhat so that each year the present value of all assets could be estimated. Naturally, the same procedure would have to be used year after year. So long as procedures such as the one governing depreciation write-offs were *unchanging* and *explicit*, the restraint on price increases would prevail. Even if the true rate of return was not 10 percent as calculated but actually 13 percent because of some "cheating" in the tax code, prices could not be raised until the profit margin dropped to the lower limit. This lower limit might in reality be 9 percent rather than the proposed 6 percent. Such discrepancies merely would alter the limits. They would not contravene the proposed goal of price stability and full employment.[2]

The other aspect of PSP is that all profits in excess of a 10 percent return after normal taxes are to be paid in full to the government. In this way, companies will pass on productivity gains in the form of price reductions or wage increases or both. Why pay the excess to the government?

In table 2 we see that company A's assets in 1976 are $1,000,000. Suppose that its profit before corporate income tax was $220,000 and that corporate income taxes were $110,000, based on a 50 percent rate. That leaves a net profit after normal taxes of $110,000, which is an 11 percent return on assets. Both the profits and assets for the five-year period from 1972 through 1976 must be added together first, and then divided by one another to determine the average return for the entire five years. When this is done, the five-year average for the years 1972 through 1976 becomes 10.2 percent.

The excess profit owed the government is now only .2 percent of assets. Since this return is an average one, we must average the value of the assets for the same period and multiply that figure by .2 percent. The result is then the amount owed the government in excess of the 10 percent return. The five-year average of assets is $978,000 or ($925,000 + 975,000 + 990,000 + 1,000,000 + 1,000,000)/5. When multiplied by .2 percent, the result is $1,956, which company A must pay to the government. Although the excess profit was $10,000 in 1976 (a 10 percent return on $1 million is $100,000, and profits were $110,000), company A has to pay only $1,956 because profits are averaged over a five-year period.

A long-cherished goal of our society is the preservation and encouragement of small business. To date, we have given mainly lip service to this goal, although some antitrust suits are filed each year and we appropriate money from time to time to aid small businesses. Further help for small businesses could be achieved by incorporating into our proposal a modification permitting a different profit range for small and medium-sized companies.

Profits do provide a source of funds for expansion that smaller companies find crucial, since their ability to borrow (given their higher risk) is more limited than that of a giant corporation. With a variation in profit ranges, growth of smaller companies could be encouraged relative to larger companies.[3] This would be one more way to promote more competition, decrease concentration of corporate power, and foster innovation as well, since newer and smaller companies are often more innovative than large ones. These aims would fit in well with the desired goals of full employment and price stability.

As an illustration, small companies might be permitted a profit range of 10–14 percent and medium-sized companies a range of 8–12 percent. The giants, such as the 500 largest firms, would have to operate in the 6–10 percent range. As is apparent in these figures, the smaller companies could raise their prices even though the profit rate was still rather high (8 percent for the medium-sized companies, 10 percent for the small companies). It is not thought, however, that this modification would hinder the achievement of price stability, since the smaller companies have to be more competitive in terms of prices and do not have the impact of the largest corporations on the overall price level.

It should be noted that before any company, including those in highly concentrated markets, raises prices it must consider price elasticities of demands. By this we mean that the company must estimate the response of the buyers to a higher price and the price response of competing firms as well. If the predicted response of buyers and competitors is regarded as negligible, then prices are increased as long as profits are being increased. If a substantial reduction in sales is expected following a price increase, companies will decide not to raise prices given that such a move would reduce profits.

There is a bias in the PSP scheme in favor of wages over profits. Wages can increase but profits must fall before the prices can be increased to restore the former profit level. This bias, however, will help improve the distribution of income, which has not changed significantly in recent years:

> In 1947 the bottom fifth of the families received five per cent

of aggregate income; in 1966 the bottom fifth of the families received five per cent of aggregate income.[4]

Here is how a change in that income distribution might come about. Profits become dividends and capital gains. These are concentrated in the hands of a relatively small number of stockholders. If ceilings were established on that part of the income of these wealthy stockholders, then their income might not rise so quickly as it does now. If wages rose steadily and prices did not increase so rapidly as in the usual current pattern, then wage earners might earn a larger share of the total income. Wage earners tend to be the lower and middle income groups. Even those in the lower income groups, without substantial wage increases, would benefit from the price stability relative to the present trends.

Reasonably competitive industries such as agriculture, textiles, and furniture would be taxed in the same way; that is, all profits in excess of a five-year average return of 10 percent would be owed the government. Prices in such industries, however, would be free to fluctuate in a manner determined by the market. In such industries, price and wage increases are restrained by the market. Price agreements are not so easily arranged, owing to the abundance of firms. Unions are likewise not so powerful. The freer entry into such industry tends to keep profits from being excessive. (Even in the agricultural sector, however, there is increasing evidence of economic concentration.) Competitive industries, consequently, are not the industries that initiate the cost-push inflation. The 10 percent profit margin after normal taxes, however, could serve as a guidepost for prices and wages in such an industry, since all profits above that level would be taxed at the rate of 100 percent.

PSP may raise some questions regarding the home construction industry, where costs have risen extremely rapidly. Given our growing population and its concentration in urban areas where land is scarce, land prices rise rapidly. Rising land values are an important factor in the increase in housing prices. In order to curb the growth of this type of inflation, we propose that all capital gains from rising land values in excess of an average annual increase of

10 percent be taxed as excess profits, regardless of the time period involved—whether it is one day or fifteen years. A period of six months, then, would allow an increase of about 4.8 percent, as this results in an annual rate of 10 percent.

For example, at a 10 percent annual rate of return, the value of an investment will double in approximately seven years. When property is sold, if the value has more than doubled in each seven-year period, then this excess would have to be paid the government. Suppose a person bought some land for $5,000. At the end of fourteen years, let us assume that he sold it for $30,000. According to the 10 percent return, that land would double at the end of seven years for a total of $10,000, and double again at the end of the next seven years for a total of $20,000. The owner, nevertheless, receives $30,000 for its sale. He will pay the usual capital gains tax on the sale of the property for $30,000. The usual capital gains tax would be considered the normal tax payment. After this payment, however, anything left to the owner in excess of $20,000 would have to be paid in full to the government. The $20,000 is considered the maximum after-tax return permitted for this fourteen-year period. Everything else is regarded as excess profit. This same rule could be applied to all real property and would facilitate land-use planning rather than interfering with it. Such a scheme would undoubtedly retard the increase in price of such scarce factors as land and would hence control this component of inflation.

Profit Constraint in Perspective

In Europe, where governments also struggle with efforts to control inflation rates while maintaining full employment, price control is usually called an "incomes" policy, as it includes the effort to moderate increases not only in wages and prices but in profits as well. Orthodox European economics thus is clearly interventionist economics, for the European expects his government not only to provide full employment at a rate of 3 percent or lower unemployment, but also to see that standards of living rise for all workers systematically and that the costs of inflation are equitably distributed among the population. Intervention programs may vary among governments and parties, but no parliamentary democracy could survive without attempting to achieve these aims. The American economist is aware of these facts, which indicate that there is nothing about economic theory which precludes such an incomes policy. Thus the discernible difference between American and European orthodox economics must be due either to empirical differences or to ideological bias.

Historically, the United State Congress, in spite of economists' timidity, has systematically attempted to control profits, the economic power associated with economic concentration, and the resulting disproportionately large incomes. Our public utility legislation, initiated on a state and local level, is a traditional type of profit control. In this century, every war except the Vietnam War has added to our experience with laws controlling war-bulged profits through excess profits taxes. Indeed, the pursuit of national policy *without* profit control was politically impossible to an American wartime Congress. The nation's most recent experience with profit control was during the Korean War, when the economic and political issues were thoroughly debated in Congress and in the professional journals. The principle was so acceptable in Congress that an excess profits tax bill introduced in the House of Representatives on June 20, 1950 was out of the committee two days later and passed by the House on June 29.

Congress successfully wrestled with issues such as deciding what was the appropriate base to be considered "normal," whether the law was to cover unincorporated as well as corporate business, what was to be done regarding "strategic products," new corporations, and depressed industries, and so on. Economists, debating as usual, were nevertheless ready to accept social policy and to do their job of analyzing its effects, including its inequities. Several economists thought that such laws encouraged growth in terms of research and development expenditures, employment rolls, and, possibly, increasing expenditures of individual firms. Most felt that special considerations had to be provided for new firms and the expansion of firms needing equity financing. The point is that American economists faced with the reality of profit constraints are technically able to help Congress tailor a workable program.

Without concurrent price control of some sort, most economists would agree that excess profits taxes will be, as they were later in the fifties, simply passed on to the consumer.[5] Consequently, most economists associate successful efforts at controlling profits through taxation with wartime experience which coincided with general price control.

Thus, given our historical experience, the orthodox American economist would probably agree that excess profits taxes are politically unavoidable in wartime, and that they have a reasonable chance of success only if they are related to some form of wage and price control—a sort of American "incomes" policy, although it is not called by that name. This conjures up the specter of an economy rigidly controlled by a huge bureaucracy. Undoubtedly, such rigidity convinces many economists, even those quite willing to engage in reformist intervention, that profit controls are a last resort, justifiable only in wartime. Yet even the Nixon administration used profits as a guideline for price increases, and England is now doing something similar.

The Permanence of a Large Governmental Role in the American Economy

Why are profits controls considered justifiable, indeed mandatory, during wartime? Various reasons may be given, but since the Korean War was not a total war effort, the most reasonable economic view is surely that during a war period, the United States government is assuming a crucial role in the market economy. As a huge buyer, providing extraordinary demand, the government acts in such a way that the economy is so dislocated as not to be functioning "fairly" in the market sense. It is our argument that these conditions have become the rule in the American economy, rather than the wartime exception. Several changes have occurred or are occurring which justify this view.

(1) Demand is growing most rapidly in the area of services,

many of which, such as recreation, education, and mass local transportation, are local or federal responsibilities.

(2) Government is assuming the role of "employer of last resort." It is becoming responsible for providing sufficient numbers and kinds of jobs to supplement private employment and provide full employment at some defined level. This is in an experimental stage in the Emergency Employment Act of 1971 and other federal-local revenue-sharing programs aimed at providing federal funds for local services such as law enforcement and summer youth programs. It is probably an inevitable outcome of the Full Employment Act of 1946, in which the federal government assumed some responsibility for full employment. Coupled with the newer principle of revenue sharing, both these principles have extensive bipartisan support. It would be foolish to believe that such fundamental changes in governmental responsibilities have no further implications for the control of their effects.

(3) Government is assuming the role of "lender of last resort" to certain companies, as is seen most notably in the controversial bill regarding the Lockheed loan. This action has a deep historical root in the Reconstruction Finance Corporation of depression days, which was a Hoover rather than a Roosevelt innovation.

(4) Government assumed the role of "lender of last resort" to banks during the thirties. This was surely a predictable outcome of the establishment of the Federal Reserve system.

What all this means is that government, both federal and local, is now guaranteeing jobs and profitability to individuals, banks, and businessmen, and that the process of change has been going on too long for it to be reversed as Friedman thinks it should be. Perhaps it began when the Supreme Court recognized the corporation as a legal individual with constitutional rights and protection. Rights and protection inevitably imply regulation, it would seem. Labor leaders were stunned to learn that this was so, and for them, surely, the period of uncontrolled behavior was briefer than that allowed the corporation. In 1936 unions and their members gained the protection of the Wagner Act; in 1947 they began to feel the effects of regulation of their own internal decisions and activities.

Thus we conclude that the debate among American economists

must shift from the issue of whether or not to intervene, to the issue of putting away the toys of laissez-faire and participating in the shaping of the economic policies of the American government. In this context, historical, theoretical, and empirical, we have put forward the Profit Stabilization Principle as one possible approach.

The reader has no doubt thought of numerous objections to this scheme. But at the same time many benefits are apparent. The following chapter examines some of these objections, which may be referred to as costs, as well as some of the benefits of PSP.

3
Costs and Benefits of the Profit Stabilization Principle

Our plan, referred to as PSP, has some costs as well as benefits. We now propose to analyze in some detail the criticisms that can be made of our scheme, as well as the benefits. The question then becomes one of whether or not the benefits exceed the costs.

Costs

Since profits are a key source of funds for expansion of a company, restrictions on profit conceivably could restrain growth. For example, in 1968 undistributed profits were $24.2 billion and capital consumption allowances were $46.8 billion. It is obvious that savings generated internally by the corporations provide enormous funds for industrial expansion.[1] See table 3 for profits and capital consumption allowances since 1950.

It is possible that certain types of frenzied growth would be affected by PSP in the following way. Some companies reinvest a large portion of their high profits. This has undoubtedly been a factor in their rapid growth. A limitation on profits might force such companies to seek funds in the money market, where availability of credit may be limited as a result of monetary policy. For this reason, there might be less growth. The more stable type of expansion would continue, however, as PSP permits a continuation of the long-run average profit rate of 8 to 10 percent. And smaller companies would be permitted a higher maximum profit as noted earlier. They could therefore use their funds for accelerated growth.

31

TABLE 3

Corporate Profits and Capital Consumption Allowances, 1950-72 (in billions)

Year	Corporate Profits After Taxes — Total	Dividends	Undistributed Profits	Corporate Capital Consumption Allowance	Undistributed Profits plus Capital Consumption
1950	24.9	8.8	16.0	8.8	24.8
1951	21.6	8.6	13.0	10.3	23.3
1952	19.6	8.6	11.0	11.5	22.5
1953	20.4	8.9	11.5	13.2	24.7
1954	20.6	9.3	11.3	15.0	26.3
1955	27.0	10.5	16.5	17.4	33.9
1956	27.2	11.3	15.9	18.9	34.8
1957	26.0	11.7	14.2	20.8	35.0
1958	22.3	11.6	10.8	22.0	32.8
1959	28.5	12.6	15.9	23.5	39.4
1960	26.7	13.4	13.2	24.9	38.1
1961	27.2	13.8	13.5	26.2	39.7
1962	31.2	15.2	16.0	30.1	46.1
1963	33.1	16.5	16.6	31.8	48.4
1964	38.4	17.8	20.6	33.9	54.5
1965	46.5	19.8	26.7	36.4	63.1
1966	49.9	20.8	29.1	39.5	68.6
1967	46.6	21.4	25.3	43.0	68.3
1968	47.8	23.6	24.2	46.8	71.0
1969	44.8	24.3	20.5	51.9	72.4
1970	40.2	24.8	15.4	55.2	70.6
1971	45.9	25.4	20.5	60.3	80.8
1972p	52.6	26.4	26.3	67.7	94.0

Source: *Economic Report of the President,* January 1973, Table C-73
p: preliminary

(Growth and the problems associated with it are discussed in more detail in chapter 5.)

One of the current problems in our economy is the fluctuation in growth from one extreme to the other. For example, our economy may have a 2 percent growth rate for a year or two, followed by gradual expansion and then a high rate of 5 or 6 percent in spurts.

Such fluctuations necessitate compensatory fiscal and monetary policies, which sometimes overshoot their mark and create further problems, such as more inflation or a recession with zero growth. A more stable, if less spectacular, growth may result from our proposed PSP. That is, growth may also be stabilized along with profits. And in the long run a steadier growth can achieve just as much at a constant lower rate as one with alternating extremes.

For example, a 5 percent annual rate of increase over a twelve-year period yields an increment of 79.6 percent. A rate that alternates between 7 and 3 percent yields an increment of 79.2 percent. The two increments at the end of the twelve-year period are virtually identical, but the former increment is achieved with more stability. If this were true of the economy as a whole, then such stability would imply a reduction in the swings between unemployment and full employment and in all the social miseries ensuing from these swings.[2]

One objection to PSP that comes readily to mind is that prices and profits would no longer serve their function of allocating resources. That is, when prices and profits rise in a certain sector, this is the signal to increase production in that sector. Our reply is that whereas prices of key products are not permitted much fluctuation under this scheme, profits could still move in a range between a negative return and the maximum of 10 percent which would then function as a limit on man's acquisitive spirit.

Another question that arises is whether or not a tax on profits above a certain level encourages inefficiency. It is charged that there will be no incentive for improvements in productivity. Under PSP, however, profits would generally fluctuate between 6 and 10 percent. A rate of 9 percent is vastly superior to one of 6 percent. If the company's assets are $1 million, a 6 percent return represents $60,000 in profit. A 9 percent return represents $90,000 for an increase of $30,000, or 50 percent. Therefore, a company has every inducement to innovate and keep the return as close to 10 percent as possible. It may be true, however, that the 100 percent tax on profits above a net return of 10 percent would encourage companies operating near that upper limit to engage in unnecessary expenditures via advertising, promotion of executives, padding of expense accounts, and the like. Such companies would seek to reduce their profits by raising their "costs."

To some extent, this is already being done, and will no doubt continue. But it should be recalled that profits for some individual years can exceed 10 percent in order to offset other less lucrative years in the five-year average. The padding of costs would be likely to occur only if the five-year average were approaching 10 percent. But even then, the unions would be alert to any unusual increase in expenditures of this type and would demand compensatory wage increases. On the other hand, for the same reason a company might spend more time and energy on improving public service aspects of its operation, such as pollution control, racial integration among employees, nurseries for children, and higher quality products. Also, short-run cost cutting in the form of layoffs of employees might be reduced in those companies that have high profit rates. Instead, companies could use these employees in a positive and innovative manner in these types of public service programs.

Viewed in this manner, the objection that companies may build up costs is actually an asset. Under present tax rules, a corporation may aggravate the unemployment problem by really unnecessary layoffs from among the regular work force. By doing so the corporation is shifting the burden of the maintenance of its needed labor force to unemployment compensation, welfare rolls, and the private savings of the laborers. Present tax law then allows the corporation to retain the "savings," perhaps in the form of blown-up capital consumption allowances. This may be labeled "cost efficiency" by the individual corporation, but it represents "resource inefficiency" to the society as a whole.

Without requiring that corporations maintain their own labor force requirements, PSP would at least encourage firms to do so. While PSP might at first seem to be fostering inefficiency at the firm level, actually it is simultaneously attacking both the problem of unemployment and that of prices, thus attaining a higher level of efficiency for the economy as a whole.[3]

As for productivity, worldwide competition is still the best spur for that. Present tax laws err in promoting profit retention by domestic corporations without requiring any kind of social performance, such as investment, on the part of corporations. As Keynes taught us, savers and investors are different people, even when they are corporations.

It is not thought that even "employee hoarding" would create excessive demand since, in the authors' opinion, total demand in our economy, whether private or public, is generally insufficient to provide the employment our labor force wants and seeks.

It might be argued that PSP would encourage cheating on tax returns, or that it would be effective only if businesses were honest. PSP, however, depends on honesty no more than does the present tax system. A significant improvement in any system would result from tax returns that are open to the public. This we recommend.

Unincorporated businesses are excluded from consideration in PSP because these businesses generally are not large enough to have a substantial inflationary impact and because their owners would be heavily taxed according to the federal revenue tax rates, if a few loopholes were plugged. The Excess Profits Tax Act of 1950 did not cover unincorporated businesses either.

Some people may wonder how the scheme will affect service industries, which have a higher rate of inflation than most other industries. Between 1960 and 1969, the consumer price index of all items rose by 24 percent, but the index of all services rose by 35 percent. Our reply is that if hospitals and other service groups, such as recreational establishments, are incorporated, they will be subject to the same profit and price restraints as every other corporation. Furthermore, the wage increases demanded in this sector are very much influenced by the pattern set in other industries where the unions are well organized. If the wage demands of the large unions are restrained by the limitation on profit and price increases, then that restraint should carry over into the service sector of the economy.

A complication may ensue for a large company operating in a variety of markets. As a result of the conglomerate movement, many large corporations are highly diversified. Consequently, some of their products may be sold in competitive markets and others in monopolistic markets. In the latter case, prices can rise only if profits drop below a minimum level. In the former case, where there is more competition, supply and demand will determine prices in the market. The question arises, then, of how profit and price constraints will operate with a multiproduct company.

The simplest approach seems the best one here, and it is as follows: The upper limit on profit will apply to the overall profits generated by any and all phases of the company's business activity. Any profits beyond the 10 percent maximum will be paid in full to the government. Prices of products in competitive markets can rise and fall as the market determines. Prices of products in highly concentrated sectors will be constrained by the rate of profit. Only if the overall profit rate for the company drops below the established minimum of 6 percent can the prices of goods be increased in the highly concentrated markets. Table 4 illustrates this point. The data are taken from table 2 in chapter 2 and refer to the largest corporations with a range of 6–10 percent. Assume company A sells two products—X and Y. Y is sold in a concentrated market, and X is sold in a competitive market.

TABLE 4

Overall Profit Rate: Company A

Five-Year Period	Average Profit Percentage
1975-79	6.7
1976-80	5.7
1977-81	5.2
1978-82	6.8

At the end of 1980, given the overall profit rate of 5.7 percent, company A can raise the price of product Y, since it is sold in a concentrated market. Product X can fluctuate according to market conditions, since it is sold in a competitive market. At the end of 1982, product Y's price is frozen because of the overall profit rate of 6.8 percent. Even if product Y (monopolistic market) proves to be an unprofitable line of goods, its price cannot be raised so long as product X's profitability is enough to generate an overall return of more than 6 percent.

If overall profits drop below the minimum, thus meeting the conditions for a price increase, a diversified company can use its discretion as to which product price should be raised. The company can then raise prices on one or all of the products sold in monopolistic markets. It is probable that a company would raise prices on those products that are very "price inelastic," that is, where the quantity of sales would not decline much if the price was increased. Price inelasticity could occur if the product was highly specialized and there were no adequate substitutes for it.

It is very possible that at certain times the company may choose not to raise prices even if profits are low, provided its competitors are not permitted a price increase. Suppose the competitors' profits are too high to allow them a price increase and that there are only three producers. In such a case, if one company raises its prices and the others maintain their current prices, the company that charges more may lose a substantial portion of its sales to its competitors. This is illustrated in table 5. The figures for company A come from table 2, and we add two more hypothetical companies.

TABLE 5

Net Profit as Percentage of Assets

Five-Year Periods	Company A	Company B	Company C
1975-79	6.7	7.0	8.0
1976-80	5.7	6.5	7.0
1977-81	5.2	6.0	6.5
1978-82	6.8	6.5	6.0

In table 5 we see that only company A could raise its prices at the end of 1980 and 1981. The profits of company B and company C never fell below 6 percent; therefore, they could not raise their prices. If these three companies had engaged in price collusion in the past, that would no longer be permitted under PSP. That is, if company A raised its prices, companies B and C could not follow. For this reason it is possible that company A may choose not to raise its prices in order not to lose sales to its competitors. In this sense, it is possible that price competition would be enhanced under PSP.

Given the profit and price constraints under PSP, the least profitable firm in the table above, company A, may eventually go out of business. This reduces the number of firms to two, and the industry approaches a monopoly. The result for the public of such a trend, however, is not higher prices but probably prices that are lower than they might otherwise have been. This tendency may well prevail, since only the more efficient firms could survive under PSP. But they could not use their monopolistic position in the market to exploit customers as they now do. The proposed price and profit constraints would prevent that.

One possible outcome of this scheme is that more profitable companies may be encouraged to merge with less profitable companies. An incentive for this type of behavior already exists for tax purposes. The conglomerate might be even more attractive than it is now, however, for an additional reason. If no price on any product of a corporation can be raised until the overall profit level has dropped below 6 percent, then a corporation may find it desirable to merge with less profitable ones. In this way, it could raise some of its prices and also its profits, provided its return after merger dropped below 6 percent. It will be recalled that our proposal permits prices to be raised sufficiently to produce a five-year average return of 10 percent.

There are restraints on such behavior, however. A merger with a company showing losses in order to reduce the profit rate below 6 percent would have to result in an average rate of less than 6 percent for a *five-year period*. This five-year period should be designated as the time following the merger. So it would not simply be a matter of merging and then raising prices the next year. An average rate of less than 6 percent for a five-year period after the merger would be a prerequisite for a price increase. This is a long period, and certainly a company would not plan such a move without exploring many other possibilities for raising its revenue. An additional restraint is that a company might not choose to raise prices, even when permitted to do so under the scheme. As explained earlier, one company would probably not raise its prices unless other companies raised theirs simultaneously. And the ability of the other companies to raise their prices would depend on their rates of profit. There is no reason to expect that all of them would have identical profits.

It might be argued that PSP would encourage flights of capital from the United States to foreign countries. This could happen if the United States had a profit maximum and other countries did not. PSP would not encourage production in foreign countries by American-owned companies, however. The reason is that foreign subsidiaries would be treated in the same manner as domestic ones. For example, if an American company earned a net profit of $1 million on a branch in Japan, normal taxes would be deducted as usual. In this case, however, normal taxes would include not only United States corporate income tax but also all taxes paid to the Japanese government. These profits would then be added to those of the domestic branches of the same American company. PSP would then apply to the aggregate.

Under some circumstances, however, PSP might encourage American companies to incorporate in foreign countries. For example, a company now incorporated in New York might decide to incorporate in Switzerland. If it continued producing in the United States, however, PSP would still apply because branches of foreign-owned companies operating in the United States would be subject to PSP. Such would not be the case if the company incorporated abroad and also produced abroad. In that case, the company would literally have abandoned the United States, even though the stockholders might be continuing to live there and receive their dividends there. An incentive for incorporating and producing abroad might be the differential between the maximum rate of 10 percent in the United States and a supposedly unlimited rate in a foreign country.

Profit differentials already exist between countries; that is one reason why Americans buy stocks abroad. There are many risks involved, however, and these can act as deterrents. Nevertheless, if this type of operation became extensive and resulted in increased outflows of capital to foreign countries, then two alternatives would be open to the United States. Either we would have to impose controls on transfers of money out of the country or we would have to negotiate a plan similar to PSP in other countries where we allowed capital exports. Such a plan might be as appealing to foreigners as to Americans. It might make foreign capital more welcome than it now is. This alternative is discussed in more detail in chapter 5.

These proposed changes would have a profound effect on future fiscal and monetary policies—mostly beneficial. In our next section we wish to explore the benefits of PSP. But we would like to close this section with an apt quotation from John Maynard Keynes's *Essays in Persuasion:*

> When the accumulation of wealth is no longer of high social importance, there will be great changes in the code of morals. We shall be able to rid ourselves of many of the pseudo-moral principles which have hag-ridden us for two hundred years, by which we have exalted some of the most distasteful of human qualities into the position of the highest virtues. We shall be able to afford to dare to assess the money-motive at its true value. The love of money as a possession—as distinguished from the love of money as a means to the enjoyments and realities of life—will be recognized for what it is, a somewhat disgusting morbidity, one of those semi-criminal, semi-pathological propensities, which one hands over with a shudder to the specialists in mental disease. All kinds of social customs and economic practices, affecting the distribution of wealth and of economic rewards and penalties, which we now maintain at all costs, however distasteful and unjust they may be in themselves, because they are tremendously useful in promoting the accumulation of capital, we shall then be free, at last, to discard.[4]

Benefits

The major direct benefit of the Profit Stabilization Principle is that, for the first time, private initiative would be enlisted in the effort to combat cost-push inflation. That is to say, under those rules, business and unions would find it advantageous to stabilize their own profits and wages, and elaborate enforcement procedures would not be necessary. Price increases would be the result of an objective criterion—the profit rate—and not the result of a decision by a price control board which has to weigh political factors. There is an analogy here to the complex and varied criteria for welfare versus the simplified program known as the negative income tax or family assistance plan as proposed by President Nixon.

The indirect advantages of this removal of private incentive for cost-push inflation would accrue to nearly everyone in the economy. Long-range planning by individuals, businesses, and governments would be increasingly possible. A major consideration is that PSP would provide *adjustment time* for more effective fiscal and monetary policies.

Fiscal Policy

Fiscal policy refers to the government's spending and taxing programs. Tax cuts and government spending can succeed in restoring aggregate demand if their magnitude is sufficient. This was well demonstrated during the 1960s by the tax cuts in the early years and the Vietnam War spending in the later years. We certainly achieved full employment, but it was accompanied by severe inflation. Excessive demand initiated the price rise, and subsequently cost-push inflation continued it. Such inflation ordinarily requires severe fiscal restraints, but if they are used, a recession will result. And even then, cost-push inflation can prevail throughout the recession. This is exactly what happened in 1970 and 1971.

Under PSP, inflationary pressure will be less severe. Consequently, more moderate fiscal measures can be used to curb any excess demand, as there will be more time allowed for a slowing down of the economy. When rampant inflation requires an urgent solution, as is sometimes the case, only a drastic curtailment of demand can possibly correct it, if anything can. Then we have a recession. It seems that with PSP we need not create a recession in order to curb excess demand. A more gradual slowing down of demand, either through higher taxes or reductions in government spending, would suffice if more time were available for intended adjustments to be made.

At times when demand is insufficient to generate full employment (defined as a 3 percent rate of unemployment), the government can reduce taxes or increase spending on desired public goods or do both without fear of inflation, given the restraints imposed by PSP. It is the fear of inflation that hinders the application of fiscal policy designed to increase demand and, consequently, employment. It has been shown that full employment can be generated via tax cuts or increased spending. With PSP, the government will be free to pursue a full employment goal without having to weigh it against the other stated goal of price stability.

Many people argue persuasively that we do not have sufficient public investment in such activities as transportation, housing, health, pollution control, and education. Spending in these areas could be increased with beneficial effects for all in an effort to maintain full employment, without fear of inflation. The government would no longer be hampered by the deep conflict between full employment and price stability that deters so much planning and action in the sphere of social or public investments.

It follows that if the economic activity is more stable, then the government's income (tax revenue) is also more stable. This is another advantage for the government in planning programs.

At present, the income tax offers some countercyclical advantages—when income goes up, taxes rise. Some taxes rise by an even greater proportion than income. PSP would reinforce this. As incomes rise, so do profits. And should profits exceed the 10 percent maximum, that excess would be paid to the government. In addition, since profits would be restrained by this tax-take, fewer justifications for raising wages would exist. There would be no excess profits for the unions to share. The increases in wages would more probably be related to factors such as increases in productivity, and businessmen would be willing to grant these whenever profits were bumping against the 10 percent level. Thus, the distribution of productivity gains would improve the distribution of income, one of the important aims of fiscal policy.

With some of the pressure to concentrate on immediate problems of unemployment or inflation taken off the government, fiscal policy might be directed more toward long-term gains accruable to stable and steady economic growth. For after all, in the immediate future, economic growth is the key to unemployment—growth of the kinds of jobs and the kinds of goods and services (not all material) that will allow this nation to achieve the goal of full employment. The difference in a percentage point of employment today represents approximately 800,000 jobs and their resultant services. PSP will help to provide a climate for the free economy where private and public decisions can be directed toward the achievement of these worthy objectives. In the longer run, other factors must be considered, and these will be discussed later.

Monetary Policy

While fiscal policy has been somewhat distorted by the requirements of short-run problems, an even more pronounced distortion has occurred in the case of monetary policy. The term "monetary policy" refers to the size and availability of the money supply and the price charged for the use of that money, i.e., the interest rate.

In a boom, monetary policy may halt price increases, but in a harsh and selective way. More than likely, it will halt production in those industries most vulnerable to rises in interest rates. This is the usual result of restrictive monetary policies. Such was the case in the sixties and was admitted by President Johnson in his economic report to the Congress. A recession in the construction industry was directly traced to Federal Reserve efforts to control inflation.

In a recession, monetary policy is not so effective, mainly because businesses tend not to invest, even at lower interest rates, if sales are not rising.

The most frustrating set of problems for monetary policy is, however, neither of these "pure" cases. It is the so-called new inflation, where prices creep up consistently in spite of substantial unemployment. This is the pattern that has faced the Federal Reserve Board since about 1956. An analysis of the trend from 1955 through 1965 (the only period of peace in the last two decades) shows that prices rose steadily for a total increase of 18 percent even with unemployment rates usually higher than 5 percent. Most economists believe this pattern is predictive of any peacetime future, so that future policy is faced with the necessity of curbing inflation at the same time that expansion in the economy is needed. Given induced spurts in economic activity, and the resultant efforts to stabilize prices, the growth of the money supply is bound to be sporadic rather than following the pattern of gradual increase that Friedman counsels as the only reasonable long-run policy. His proposal has merit, but not in an economy that is entirely unmanaged and without guidelines, and which still experiences significant fluctuations in employment.

A more steady increase in the money supply—and its counterpart, a more stable interest rate—would be most effective in maintaining full employment as an adjunct to a fiscal policy that was stimulating aggregate demand when necessary. If interest rate fluctuations were reduced, long-term public and private investment would be less sporadic than they are now. Investment plans—especially public investments such as those in schools, water, and the like—are often postponed when interest rates are deemed too high. Such intermittent investment activity, along with the fluctuations in the housing market resulting from lack of credit in the money market, contribute much to the fluctuation in economic activity which then exacerbates the unemployment problem.

The monetary benefit of PSP is that in making inflation manageable, the Federal Reserve would be permitted to implement Friedman's proposal that the money supply be increased at approximately the same rate each year. This could be effective if it were coupled with the utilization of fiscal policy to promote long-term growth at full employment levels of income. In this way we could achieve as much economic stability as is possible given our desire for freedom and the present state of our knowledge.

Long-Range Objectives

Individuals: Individuals and their plans are simply pawns under the present rules of the game. Books like *Future Shock* document the kind of individual agony that is experienced in our present society. The appeal of *Catch-22* to the young reveals their image of the kind of world they see about them. For the individual, rising prices and unemployment, or employment in tiresome or meaningless jobs, are not abstract economic phenomena but everyday threats to his well-being.

Perhaps one-third of the adults in the nation are virtually immune from those worries, because they are well insulated by family position and riches. Two-thirds, however, are threatened by one or both at some time in their lives.

For the individual, then, the dampening of inflation through PSP plus the increased probability of full employment through monetary and fiscal policy would indeed be a bonanza. Although he would find large wage increases hard to come by, he could plan for his future in a reasonable way—able to count on being able to find work when he needed it. What a different situation from the present one in which many people are becoming nearly apathetic from continual frustration!

Businesses: In the same way, businesses could build profits on a firm basis in this more stable world. While large gains would not be possible, over the long run a business could survive and pay off well to its founders or owners. Perhaps the long-term return would be even better than it can be now, if periodic recessions could be avoided.

A hypothetical firm without PSP can earn profits without any upper limit. Let us assume that, given a recession, a firm will earn profits in a five-year period ranging from 2 to 15 percent. If the yearly percentage rates were 15, 6, 2, 6, and 15, the average would be 8.8 percent. Under PSP the upper limit is 10 percent. Without inflation or recession, however, it would seem possible for that firm to average a similar return of 8 or 9 percent even with the upper limit of 10 percent.

Businesses could function better in a more secure environment, just as individuals could. Long-range objectives could be carried out because forecasts would be more reliable.

Government: Equally important, the governments (local, state, and federal) could maximize long-run aims. Governments, like firms and individuals, are now running on a roller-coaster economy where neither deficits nor surpluses are truly plannable at the federal level. Local and state governments are forced to delay important projects like urban transportation because of lack of funds. In short, governments, like individuals, are at the mercy of economic fluctuations. When these are minimized and the economy is more stable, the government's tax revenue will also be stabilized.

Once inflation is tamed and high levels of employment are guaranteed, then governments will be able to follow their own plans rather than merely react to the squeakiest wheel. The most important change will be that government can begin to operate on a maximization principle, at least in those activities where costs and benefits can be quantified in dollars.

The government can view itself in a manner analogous to that of a firm. That is, it can maximize its surplus—revenues minus expenditures—with respect to some sphere of its activity. Profit or surplus is maximized at the point where additional (marginal) revenue equals additional costs. When the government's expenditures on some project lead to additional tax receipts, the government can expand to the point where additional tax revenue equals additional expenditures. At this level of operation the government has maximized its surplus. Then the total surplus in this activity can be used to finance public activities which generate little or no tax revenues, such as providing food and clothing for the needy or subsidizing the performing arts and recreation.

If the government were to maximize its budget surplus solely to generate a surplus or to provide funds for paying off the public debt, the effect on the economy would be a depressing one. By taking in more money than it spends, the government would be reducing the total demand for goods and services. Production would tend to fall if the government saved some of its income via a budget surplus without offsetting its savings by spending in the private sector; the surplus generated in one sphere should be spent to cover a deficit in another activity. This is also the way a large corporation operates. Expenditures on one type of activity which are not directly profitable are made with funds generated in a more profitable line of business.

Furthermore, when the government engages in an activity, it must take into account not only all direct expenditures, but also indirect expenditures which may rise or fall as a consequence of its action. An example is the reduction in crime and welfare costs resulting from a program to raise the educational level of the poor. It must take into account not only all tax revenue resulting directly from such activity but any revenue accruing indirectly as well.

The maximization process here means that the government should invest in such programs up to the point where additional tax revenue (direct and indirect) equals additional expenditures (direct and indirect) on the project. In this way, government spending can become more rational. Surpluses generated in these activities can be used to finance deficits in others.

A useful illustration of this approach is given by a presentation of the results of a study the authors made regarding the costs and benefits of education when a group of poorly educated young men raise their educational attainment to that of the national average.[5] In measuring the benefits of education in this study, we found that the indirect benefits of higher education resulted in a reduction of welfare costs, crime costs, and costs of unemployment insurance. And the reduction in these costs amounted to half of the increased educational costs. Furthermore, these indirect benefits are only a part of the total benefits. Ultimately, the additional tax revenue which we refer to as a direct benefit amounts to 1½ times the educational costs. The time element has to be considered, however, and for more details we refer the readers to our study. In any case, such a program is ultimately self-financing. This means that it can make some people better off without making others worse off.

What about other public projects such as transportation, birth control, drug control, mental health? Our point is that government planning can be made more rational by concentrating government investment in areas that bring about returns in revenue or reduced costs or both. Furthermore, some of the services people want can be provided in this way, and some of the jobs they want can thus be created.

Role of PSP

Obviously, PSP is only the first step in reconstructing a chaotic economy toward order. Other changes, such as the elimination of regressive taxes and the plugging of tax loopholes for the rich, would bring great benefits. But adoption of PSP would signal our willingness to provide rules which, given normal private incentives toward self-benefit, would move individuals and businesses toward common social goals.

In one sense, PSP would make regulated companies of all private businesses, but in another sense business would be *virtually without regulation*, because all that is involved in PSP is an upper limit on profits and the requirement that price increases depend on profit rates. Anything else would be up to the firm or individual. Companies could learn to live with rules that provided stability and other social values for the rest of us.[6] This is the way the American economy was always intended to operate. American capitalism should not be a system primarily intended to benefit the rich and to make them powerful enough to subvert the democratic will of the people. American capitalism and its system of private initiative was intended to benefit us all and to *avoid* systematic corruption of the government. In this perspective, PSP offers political as well as economic advantages.

Freedom and Control

The basic crisis in American society today is not, after all, an economic one. One can point out too many achievements in that sector to believe that we are near economic collapse. The basic crisis is centered on the problem of freedom and control. We are trying to attain a higher level of performance not only in the economy, but in every other facet of our social lives as well. We want our cities not just to be economically livable, in the sense that people have jobs and enough to eat and wear. We also want them to be places where people can live together willingly and joyously, free from more than the minimum necessary coercion.

The outline of the solution to this crisis is deceptively simple for some of us. We must place as few of the social variables as possible under political control, leaving most decisions to individuals motivated by the desire for what is best for them, i.e., under control by more random social variables.

Whereas in earlier times we were threatened by despotic kings and governments, we are now more threatened by behavior modifiers who would use their expertise either in the service of well-meaning governments or to ends which they themselves determine. Charlatans as well as scholars can read B. F. Skinner to learn how this can be done.

PSP does offer a way to change the rules of the major game of the economy. Perhaps it might even be said that PSP would represent the manipulation of the motivating signals in this game—the prices, profits, and so on which have heretofore not been subject to manipulation. But this would be a less-than-half-truth. For every schoolchild must know by now that government guarantees some prices (farm), engages in collusion to maintain other prices at artificially high levels (rail and air), and restrains others from rising (utilities). That schoolchild may also know that for some industries (oil and extractive) a large share of the cost of doing private business for profit is tax-sheltered in comparison with the costs of other industries. Many businesses (cooperatives) have tax rules of their own. Still others (air and merchant marine) receive direct subsidies in the form of government investment. And the list could go on and on.

It is plain that manipulation is already here, including a virtual tax subsidy for businesses that invest abroad rather than in the United States where they would hire our own unemployed. So PSP would not be new in the sense of introducing manipulation where there is none. Let us concede that it would be manipulation, nevertheless. The point is that it would be manipulation merely of signals, not of persons. Direct controls on prices and wages would be unnecessary.

Most Americans either have not realized or have preferred not to notice what the emergence of gigantic profits has done to the structure of their government and hence to their own freedom. Perhaps one instance—a tiny one after all—may suffice. Standard Oil of California contributed an admitted $75,000 to defeat (successfully) an initiative proposition which would have subjected the development of offshore oil resources in the state of California to greater public control. Two other oil companies admitted to having made smaller contributions in the same campaign. Is that any argument for leaving these large sums in the hands of businesses which are responsible to no one, not even to their dispersed and apathetic stockholders? PSP would cut down this power a tiny bit by removing some of the surpluses that can be used for political purposes. This, we submit, is the right direction in which to move.

4

Economic Evolution and the Public Interest

What we have discussed so far is a specific proposal to reorder our economy in the near future. What we would like to examine now is a more general view of our economy and the direction in which it is headed. Then we can see how PSP fits into the evolutionary trend in our society.

It may be helpful at this point to examine the historical development of the many economic goals which have evolved in western thought and to discuss what we think to be the direction of the current trend in economic thought. For the evolutionary process is as inevitable in economic institutions as it is in the world of living organisms. This was noted long ago by such diverse economists as Adam Smith, who reacted against mercantilist reliance on balance-of-trade concepts, and Karl Marx, who reacted to what he called capitalism. Most recently Galbraith and Boulding have discussed the changing economy.

To accept this evolutionary process in institutions is to accept more than mere change. For although economic evolution may not always mean "progress," it does imply a selection process going forward in time in economic institutions, that is, in selection and in the survival of the viable and adaptable institutions. There is disagreement as to whether economic goals themselves are evolving. To accept evolution of economic goals is indeed to become hopeful—to introduce optimism into the "dismal science" of economics.

The proposal of PSP is made on the premise that the goals and institutional structures of economic systems are subject to such evolution. The form that evolution takes in economics is the result of an increased awareness of what an economic system is, what it does, and what it can do. This approach is in the spirit of Teilhard de Chardin, who defines evolution as "an ascent towards consciousness."[1] Consciousness is the awareness of ourselves and others. This increased awareness facilitates change.

The evolutionary process implies some progress in these terms, but it also connotes some conflict. A major conflict now prevailing is the one between price stability and full employment. Conflicts in this sense, rather than being considered undesirable, can be viewed as selective, competitive, or tension-inducing in the system. This conflict and tension can then drive the system on to a higher level of performance. Let us define the level of performance to mean *how well a system operates according to some criterion such as minimization of cost.* Now let us examine the various criteria that have been developed in our economy.

Level of Performance

Performance may be thought of in an engineering sense as least input for a given output in terms of real commodities. How efficient a system can be in this sense depends to a great extent on knowledge. For example, if new knowledge permits us to make a given product (or to fill a certain need) with less of something (labor, capital, or material), then the new way is more efficient than the old way. This is comparative efficiency in real terms. Underlying it is the assumption that we wish to economize: to save effort, materials, or even time.

One of the early examples of efficiency is described by Adam Smith. He had visited a pin factory and observed what he publicized as the division of labor but what really was the specialization of tasks in production. While a workman not "educated to this business . . . could scarcely . . . make one pin a day," ten men by organizing separate and distinct tasks, even with little machinery, could make twelve pounds of pins a day, with upwards of four thousand pins to the pound.[2]

The engineering test of performance—least input for given output—had a special appeal to Thorstein Veblen. In all his writings he emphasized engineering, and he thought it was the ultimate rationality in production. The engineering test could include not only private costs but all indirect costs to society as well, such as pollution of air and water, noise, and other external effects.

The more traditional economists hold out for another test of performance which they refer to as *cost efficiency*. Taking prices as given (and this is an important point), they argue that only the monetary cost of inputs should determine whether one process is more efficient than another, or performs better than another. This process is usually viewed from the standpoint of the private producer seeking to maximize profits exclusive of any indirect costs to others. Now, however, more and more economists are including the external effects on the environment in their cost analysis.[3]

These two ways of measuring performance—engineering efficiency and cost efficiency—are comparative methods. Engineering efficiency tells us that the best performance is that which saves most in terms of *all* effort, material, and time for society as a whole. Cost efficiency tells us that the best performance is the one which is least costly in money terms with respect to some time factor and usually from the standpoint of the private profit maximizer.

Veblen asserted that cost efficiency resulting in profit maximization could not approximate engineering efficiency decisions. Not only are external effects often excluded, but prices are distorted by monopolies. His charges were based partly on the investigation of trusts by the Federal Industrial Commission of 1901. Monopolies and trusts establish pricing structures different from those of a competitive market. Consequently, money costs do not always reflect real costs. Even when money costs are the same as real costs, however, not all costs are counted in some industries. A private industry that pollutes air or water may not be required to clean up the mess, and society suffers as a consequence. There is, therefore, a cost to society which is not included in the private costs to the firm. As a result, more of this product may be manufactured and consumed than would be the case if the private sector had to correct the pollution damage, and hence raise the price of its product.

Another test of performance utilized by economists is to ask whether or not an economic system is operating at something near its *capabilities*. For the whole economy, this is usually taken to mean full employment of labor and existing capital. In a given industry, it is a measure of whether it has "excess capacity." In either case, unit costs are reduced substantially as the economy moves close to, but not beyond, its "intended" or natural (or cultural, as in the case of labor) capacity. One of the reasons for the reduction in costs is that total fixed costs are spread over more units at full capacity. Assume that fixed costs—buildings and equipment—amount to $1 million annually. When the equipment is operating at 50 percent of capacity, 1,000 units are produced. The fixed costs, therefore, are $1,000 per unit. When the equipment operates at full capacity, 2,000 units are produced, and fixed costs per unit are only $500. This is a tremendous advantage in terms of profit for any company, and additional variable costs do not immediately offset it.

A standard much discussed today is the *rate of growth*. The higher the rate, the better the performance. If an economy is operating at full capacity, a high rate of growth usually follows.

One last measure of performance has an even longer history of public (political) acceptance than does full employment. This is the goal of stability, usually stated in economic terms as *price stability*. Fifty-nine years ago the Federal Reserve system was created to achieve this goal and others. At that time the Federal Reserve system was given the power to change the rates at which member banks could rediscount loans in order to extend the member banks' lending power. By raising the rate, the Federal Reserve boards could tighten up on credit expansion. Also, the Federal Reserve boards (FRBs) could raise or lower the minimum reserve requirements, making credit tighter or easier, as they wished. Only incidentally was the system given the power to conduct open market operations in federal government bonds, that is, buying and selling government bonds. This was thought to be an important, convenient power. But even with a small federal debt of $16.5 billion in the twenties, the Federal Reserve Board of the powerful New York bank began to develop the technique of using open market operations to affect credit. Of course, when the debt expanded during the depression to $44.8 billion by 1940 and later to $252.5 billion by 1945, open market operations became one of the most sensitive and powerful weapons of the system. And the Federal Reserve system did attempt over those years to maintain price stability through use of these powers.

The reader must already see that the various performance tests can be in conflict—that is, engineering and cost efficiency (profit maximization) tests may conflict. Even if they are in harmony, capacity production may fall short of a satisfactory level. This is exactly what has occurred in the American economy. It is the rule rather than the exception today. Unemployment was persistent during the 1920s, as is reflected in the rising local expenditures for unemployment relief.[4] Some industries, particularly agriculture and coal mining, were labeled "sick" because of inadequate demand for their products. In the 1930s, unemployment reached a catastrophic one-fourth of the labor force. Today, industries associated with unused labor and capital are, intermittently, construction, textiles, and war-related firms, to name only three.

When our economy does approach full capacity, then we have a conflict with the test of price stability, as we saw earlier. Prices now rise rapidly when unemployment drops to 5 percent or less. It is the function of PSP to correct this defect in our system. It is not a simple problem of bringing industries up to capacity and then sitting back to watch the system run smoothly.

Conflict of Goals

To get down to brass tacks, a number of conflicts arise with respect to levels of performance. The performance tests of an economy are subject to conflict not only because of class and income differences, but also because economic development affects different people in different ways, and because there are alternative ways of measuring the level of performance, as we saw earlier. Engineering efficiency has become more prominent recently because the ecologists have become more vocal. Cost efficiency goals are still dear to the hearts of the profit-maximizers. Capacity efficiency as an overall goal was publicly adopted twenty-eight years ago in the Full Employment Act of 1946.

The Full Employment Act was first of all an act of admission that the federal government rather than the "unseen hand" theorized by Adam Smith was to be the employer of last resort. This act came out of the theories of Keynes, along with the disastrous experience of the American depression, which was much worse in terms of unemployment rates than the similar European and British depressions. The fear that the return of the enlarged World War II armed forces to civilian life would mean another serious depression also spurred Congress. To implement the purposes of the act, Congress created the Council of Economic Advisers whose members were to act as advisers to the president and to make reports to him. He was to report in turn to the Congress, which would do some of its own investigations through the Joint Economic Committee of Congress, also created by the act.

Those who are sold on cost efficiency are likely to think that price stability is not only the older and thus the more venerable goal, but also that it is the more fundamental goal. They will be willing to tolerate more unemployment to achieve it. Those who are sold on capacity efficiency will want to hold out for full employment. They will likely tolerate some inflation.

Which is best? Or rather, is a more encompassing measure of the level of performance now emerging out of the fundamental goal conflicts in our society? The authors of this book believe that this is happening and that the new goal will be an economy which serves people, and serves them well, rather than the other way around. For in each of the traditional performance measures, even that of full employment, people have been "input." They have been regarded as "resources," even as "human capital," or cogs, if you will, in the system. This book represents an effort to see how the economy would look if it were redesigned or regulated to serve people.

If we assume that the economy is to satisfy people, is there some new standard of performance to be applied? Or to put it differently, what is it that people want from our economic system that they are not now getting? Our reply is that people want more goods as well as different kinds of goods, and they want more jobs as well as different kinds of jobs. These are the goals we will examine next. Furthermore, they want these goods and jobs in a climate of price stability and a reasonable amount of personal freedom and opportunity.

More Goods

Unquestionably, some consumers want more personal economic goods than they now have. "Why don't they have these goods?" we may ask.

A. Is it because we are short of resources for producing more? While the United States is dependent on foreign sources for many basic materials, we can fulfill our needs through trade.

B. Do we have the labor and capital necessary to fulfill more of these needs? Yes, because most of the time we are operating at less than capacity. In the decades of the 1950s and the 1960s, the unemployment rate fell below 4 percent in only seven of these twenty years, and these were the years of the Korean and Vietnam wars.

C. Is it the lack of buying power, then, that separates people from goods? To a degree, yes. Twenty-seven percent of the working population still earn only eight percent of the income, so the distribution of income is not satisfactory given all the evidence of poverty in our midst.

The economy, therefore, operates at something less than a satisfactory level with respect to both production *and* distribution. That the problem includes both production and distribution should surprise no one. Economists have been arguing about this for over a hundred years. What are their arguments?

For all but a very few heretical and usually non-Marxist economists, there can never be a problem of producing too much. People always want more. We agree with this statement, but not with all its implications about unlimited growth in the future. When people are still ill-fed, ill-housed, and ill-treated, more goods and services are needed—but not forever and ever.

Many economists claim scarcity to be *the* economic problem, given the assumption that wants are insatiable. While knowledge may extend resources, knowledge also creates wants. So in general, production never can exceed "wants." Limited resources, however, can act as a constraint on production.

The long-run problem of resource exhaustion, either through using up relatively irreplaceable fossil fuels or through destructive pollution, merely acts as a deterrent to the prevailing economic view that more should be produced. (This point is discussed in more detail in the next chapter.) For an economist, production is validated by wants. To say, then, that a good or a service should not be produced for technological reasons is different from saying that the good is unwanted. The former conclusion may result from an analysis based on engineering criteria, while the latter may result solely from a change in tastes or from a price in excess of one's ability to pay. In the meantime, most people (lower and middle income groups) want and need more.

One of the clearest reasons that more goods are wanted is the simple fact that many people do not have enough money to "validate" their wants. Economists have preferred to ignore this problem in favor of emphasizing the problem of production. They generally argue that it is more important to make the pie bigger than to divide the pie differently.

Unfortunately, economists at the level of the Council of Economic Advisers have tried to sell the whole country on the idea that having more investment is better public policy than changing the distribution of income. This generally adds up to policies that cut taxes on business in the hope of getting greater private investment and more growth, rather than cutting taxes on consumers in the hope of getting more private consumption of goods, and subsequently more investment.

Probably economists argue this way partly through political pragmatism, since additions to investment are more politically acceptable than any redistribution plan, including Nixon's family assistance plan or Friedman's negative income tax. However, the additions to investment which are politically acceptable too often turn out to be the purchase or production of exotic military hardware, some of which, like the F-111, are not even suitable for the military. To avoid the latter type of expenditure, economists may recommend general tax cuts on business in the fond and often vain hope that these cuts will trigger additional private investment in nonmilitary goods.

The tax cuts proposed by the Nixon administration in 1971 were a striking example of this biased emphasis on investment to the neglect of consumption. The faster depreciation write-offs authorized by the president, along with the proposed investment tax credit, were designed to reduce corporate taxes by about $7 billion each year. Ostensibly the purpose of these tax cuts was to stimulate investment as an antirecession policy. The problem that was overlooked, but soon became apparent, was that businessmen would not expand their factories as they were already operating at something like 75 percent capacity. This meant that they could not sell what they could already produce. What was to be the point of being able to produce more? Instead, they were more likely to keep their corporate saving from reduced taxes to invest in short-term securities abroad, where interest rates were at that time higher than any domestic investment return was likely to be. Ironically, they were waiting for domestic consumption to pick up. Meanwhile, government need for government expenditures did not go down, and other taxes, such as social security taxes, went up. This would seem to discourage domestic consumption even further. Savings were huge.

With a Keynesian multiplier analysis it can be shown in the economist's own language that a tax cut on the poor can stimulate employment more than a tax cut on business when there is excess capacity and lagging consumer demand.[5] Why isn't this point argued more often?

Apparently economists still have a bias in favor of investment spending and against income redistribution via taxation changes. Not incidentally, this bias fits perfectly with the political pressures exerted by corporate and business interests. But in fairness to economists, most of them are simply more interested in investment and economic growth. As long as they are allowed to remain in the ivory tower these interests are not likely to change.

Different Goods

In addition to wanting more private economic goods, consumers in the broad sense generally want more collective or public economic goods than they now have.[6] This again is a question not only of production but of the distribution of buying power between individuals and collective groups.

Economists have not spent much time analyzing "collective demand" but clearly it falls into at least two categories. The first one is services and goods which individuals want for their own use; this category includes utilities, mails, schools, recreation sites, and roads. The second one is goods and services which businesses and corporations want as "resources," such as schools and universities from which to hire well-trained or literate employees; roads, mails, and other transport systems for carrying on business; utilities for power and water to be used in productive systems.

In the United States, the deep prejudice against government ownership has meant that even goods and services whose production is collectively regulated may be sold to the public under the guise of private ownership. Electric light and power produced on the Colorado River and sold in southern California is a case in point.

Another example is that of the synthetic rubber industry. During World War II, because of the overwhelming collective need for rubber as a war material, the United States government was forced to construct numerous rubber-producing plants. These were, however, turned over for operation to private industries, which then bought the plants after the war at bargain prices. This charade of resale was dictated not by economic necessity but by ideology.

In most other Western democracies, once industries come into collective ownership they tend to stay there. When Great Britain nationalized its steel industry, however, a part of it was denationalized—the profitable part, some say. But neither West Germany nor France shows signs of releasing its ownership rights in Volkswagen and Renault.

Perhaps the United States is the only nation where the public would permit the underwriting of capital investment *without* ownership. This we have done with the merchant marine and to a lesser extent with utilities, transportation, and agriculture. Of course, no other nation gave away its conquered lands on so large a scale as we did when we expanded westward on the continent and granted territory to railroads and homesteaders.

Despite our national prejudices, the growth of collective demand is one of the most obvious developments in the last hundred years or so. This has resulted in the American economy's becoming more "mixed" in the sense of accommodating both public and private ownership. Americans seem ready to insist on their alternatives of having things collective as well as private when and where they choose.

Projecting this collective behavior, we find no obvious drift toward the bugaboo of socialism, but rather a continued pressure for a higher level of provision of economic goods and services, whether privately or publicly owned or produced. Part of today's difficulty in obtaining more collective goods is that there is a poor, almost nonexistent, signal system by which people can indicate a preference for more public goods. Theoretically, the democratic system of one vote per person is supposed to respond to such demand. In fact, both the political party system and the pressure group system dilute any influence voters may have. How, for example, may a voter indicate his preference for a mass transit system in a world where campaign expenses are paid by highway lobbyists?

The other part of the difficulty in obtaining wanted public goods is traceable to the distribution of wealth and income. Part of the purpose of PSP is to erode the political power of monopolists and special interests in order to permit the political system to be somewhat more responsive to collective demands for goods.

More Jobs and Different Jobs

There are also certain tendencies in the labor market that foretell what people want from their economy. What appears is that people actively seek jobs which are nonexistent in most years. That is, the labor force usually surpasses in quantity the number of available jobs. People not only want more jobs, however. There are indications that they want different kinds of jobs as well.

Traditional economic theory has been based on the assumption that work is onerous to the individual and that the wage represents a payment to overcome his reluctance to make an effort. Purchased goods were assumed to give pleasure, or even life in the case of necessities, but leisure always beckoned this lazy economic man. There is undoubtedly some truth in that oversimplified picture, but we know a good bit more today about man's response to opportunities for work, income, and leisure.

For example, for whatever reason, overall labor participation rates in the United States are higher than 90 percent for all noninstitutionalized civilian males between the ages of twenty-five and fifty-four, and they approach 50 percent for females of these age groups. Among some groups, such as white males between the ages of twenty-five and forty-four, more than 97 percent are in the labor force. In addition, about 5 percent of employed persons hold two or more jobs, and most of these are young fathers with low earnings. This participation pattern may be due to economic or social coercion, but it is not the direct result of any governmental coercion. Clearly many people who could survive without working nevertheless do work regularly.

Furthermore, in spite of the dominance of large corporations over industry, the preponderance of the 5 million American business units are proprietorships. Self-employment may be a more fitting description of some of these single proprietorships, describing individuals with small amounts of capital who attempt to avoid unemployment by "buying themselves a job."

Again, the traditional approach to job creation has been through the entrepreneur who hires workmen to produce a good or a service which is demanded by a buying public. As shown in chapter 3, PSP would encourage corporations to provide regular employment for their needed work force. In fact, employer resistance to guaranteed annual wages would be significantly decreased under PSP, and the reliance of the society on unemployment compensation and welfare would likewise decrease.

Until the Emergency Employment Act, there was little concern for the creation of jobs to fit the unemployed worker or for the provision of transitional employment to prepare him for regular employment. Nor has there been any serious effort to fit jobs to qualified potential workers, particularly women. Here some adjustment in the number of hours worked per day may be necessary, especially if the woman has many family responsibilities. Why couldn't she work a six-hour shift when her children are young? Ultimately, why shouldn't everyone work a six-hour day? What about the need for more and more jobs in services such as counseling, family care, and health care? The provision of services must rise as knowledge grows and demand for them increases.

Take the demand for medical care. While the United States spent $70 billion in 1971 on medical bills, some health statistics do not reflect favorably on American health care. The usual case cited is that of life expectancy. For females in Sweden it is about seventy-seven years, while in the United States it is around seventy-four years. Infant deaths per 1,000 live births were 13 in Sweden and 21 in the United States for recent years. Between 1950 and 1971 American male longevity dropped from eleventh place in the world standings to near fortieth.

Furthermore, we cannot see ways to improve the record, largely because we are ensnared in a labor-wasting medical care system. Doctors are required by law (and sometimes greed) to do countless simple tasks like administering shots. Nurses shuffle papers, and poorly qualified persons do a substantial part of what nursing is done in hospitals. Many more trained people are needed in these areas, but many applicants to nursing schools, for example, find they must wait a semester or more for admission. Remedies for such a situation require collective action.

Part of the problem is rooted in our ideological bias. Health care is so private that in many states all the public health doctor can do is to recommend that a crippled child see the private physician he cannot afford and does not have the transportation to reach. Public hospitals usually hire immigrant doctors for nursing charity patients or to oversee beds which are unfilled. Private doctors establish their own profitable hospitals and require or encourage their patients to go there. Meanwhile citizens want better medical care, decently paying jobs, and meaningful opportunity for service, none of which is likely to come out of the present setup in this sector of our economy.

Principle by Which To Judge Performance: Public Interest

What we have tried to show is that every known performance standard has something to offer; however, if one standard is allowed to dominate, the values of the others may be lost. Hence what is needed is some principle that will enable us to determine which performance standard is to dominate which decision, or one that will show us how several standards can be achieved simultaneously. It is neither necessary nor desirable to have one standard dominating every decision.

The principle which is evolving, and which has been in evidence to some degree throughout American history, is known as "public interest." It is the meaning of public interest which has been subject to evolution, rather than the agreement that public interest is paramount. For many, the only possible definition of public interest is the sum of all individual interests.[7] Consequently, most economists would argue that any policy or standard which succeeded in making many people better off and no one worse off would be an agreeable policy. Indeed, it is hard to disagree with this idea, which is the basis for traditional welfare economics and even cost-benefit analysis. But in the exploration of the real meaning of this edict, economists have learned that it is important to introduce the idea of "diseconomies" when trying to discover whether a policy does indeed *really* make everyone well off, or alternatively, whether all of the *real* costs have been included.

It has become increasingly evident that the real trap in our thinking about the public good or public interest is that we have tended to use it passively within a laissez-faire climate of opinion to allow individuals to do whatever they wanted or whatever they felt made them well off. This was particularly true if we could not find anyone to argue persuasively for those who would be adversely affected by the individual self-interest action. Sometimes the people who would be adversely affected are not yet born; frequently they are children or poor or even old and have no powerful spokesmen. So for some of these reasons, we have come to realize that there is a public interest that includes some considerations which represent more than the sum of the immediately calculable private interests.

Historians tell us that the concept of public interest in this broader sense has always been there, for the building of roads, lighthouses, schools, parks, and even military establishments. Our forefathers knew that the individual fisherman who built a lighthouse would benefit many more people than himself. So with roads, so with schools, and so on. Yet, for at least fifty years and more likely a hundred, the subsidiary concept of cost-efficiency and profit maximization has been allowed to determine what *is* "public interest." The dictum "What's good for General Motors is good for the country" was not a new concept in the fifties when this phrase became famous. Yet the little-used public welfare clause has been in the United States Constitution all this time, and similar clauses exist in the constitutions of most states. It is the interpretation of these words that is changing as society evolves.[8]

Today, we would argue that public interest demands the achievement of an equitable economy, that is, one that serves the needs of the people and serves them well. Full employment and price stability, with increasing consumption of goods and services, both public and private, are part of what people want and need. The Profit Stabilization Principle, where greed but not initiative is harnessed, would permit the achievement of this end in an atmosphere of individual freedom.

Perhaps it is our natural fear of oppressive governments which has made us fearful that we cannot have price stability, full employment, and personal freedom. Maybe the nation needs a giant psychiatrist to explain that our ancestors' experiences with despotic kings and our contemporaries' experience with oppression by dictators need not happen here.

Americans are sufficiently imbued with a belief in individual freedom, and our institutions are sufficiently embedded in a framework guaranteeing this freedom, that we might surely shuffle some of our economic priorities without enslaving ourselves.

In the next chapter we try to delineate the public interest for the United States in the more distant future.

5
Stability Without Growth in the Eighties?

In the previous chapter we discussed the direction in which the economy is evolving. The goal that we propose is one based on public interest. Its economic meaning is full employment combined with price stability, along with increasing public and private consumption via expansion of the economy and redistribution of income. These goals are most easily achieved when a reorganization of our economy along the lines of the Profit Stabilization Principle is implemented. Increasing public consumption is to be in areas that serve the needs of the individual members of our society: housing, transportation, education, health, and the like. We exclude increased military consumption from this standard of performance.[1] In this final chapter we wish to examine the long-run implication of these goals.

An affluent economy such as ours has the time, energy, and resources to devote to the development and enhancement of the individual, if it so chooses. And we think that the popular concern about civil rights, racism, sexism, and pollution shows that we are moving in that direction. Such a propensity stems from our increased awareness. To paraphrase Teilhard de Chardin, increased self-awareness *is* evolution, and evolution relentlessly moves us to a higher plane. In a humanistic society, this means that man becomes more and more the center of the society, whose purpose is to serve each and every person well. The sole function of business, therefore, is no longer to earn profit. Its primary function is now to serve society.

In such an economy—call it a mature economy—the overwhelming issue is no longer one of obtaining sufficient food, clothing, and shelter. Each year, more and more people have satisfied their basic needs. The question now becomes one of *how* we provide ourselves with food, clothing, and shelter. What is the effect on man's spirit and body of the way in which we produce goods and services to satisfy our basic physical needs? Are people exploited and worn down by the hours of work and the environment in which they work? More recently another important question has come up. What is the effect on the physical environment of our means of production? Is nature being exploited and worn down by the intensity of production and our style of life?

In a very poor society where most people live on the borderline of subsistence, the argument that people and natural resources ought not to be cruelly exploited is not persuasive with most people. For most, life is a matter of physical survival from day to day. Such a daily challenge does not heighten man's sensitivity to his fellow man and his environment. But in a mature and rich economy like that of the United States, no one can persuasively argue that we must exploit some human beings for the benefit of others. Nor can we easily argue that the physical universe must be plundered to preserve mankind.

We are now beginning to realize that we are one with our universe. We sense our interdependence with the physical nonhuman world, and we are learning to respect it just as we are learning to respect one another. This is a consequence of our evolution over time. And evolution is equivalent to increased consciousness—the power to reflect.

Such reflection requires leisure time. It can be argued, therefore, that the freeing of more and more men by means of mechanization from laborious and prolonged work has enabled human beings to make such progress. For it is only in moments of leisure that we can be free to develop more sensitivity. We need time to reflect in order to understand people and things more fully. With more leisure available in the future, we can become closer and closer to other people and to the physical world. The culmination of all this is an increased capacity to love. And love is the fundamental impulse of the world.

Women's movements are helping us to understand that full human development is necessary for all and that such development is not solely or even primarily a function of bearing or rearing children. As human beings are being freed from grueling toil by advances in technology, mankind is also being freed from the need to procreate by the same types of technological advances that have reduced mortality. Quality now becomes paramount rather than quantity.

The consequences of human neglect in the formative years are evident in the crime, delinquency, illiteracy, drug addiction, and extreme emotional strains of many people. These consequences are largely avoidable if enough time, care, and resources are devoted to human development. And full human development, we are beginning to sense, does not mean more and more of everything. It means neither an ever-increasing population nor an ever-expanding economy. In fact, if our goal is the most perfect development possible of each person in all facets of his or her intricate personality, then it is more readily attainable if our resources and attention are devoted to a population of constant size rather than to one which is growing.

Furthermore, if our goal is the conservation of the physical world, then this too is more easily achieved if the population is stationary. In a finite universe, population and production cannot expand indefinitely. Ultimately, their growth must be curbed. Ideal human development also requires a limit on the size of the population. Therefore, we should strive for a gradual diminution in the rate of growth of population as well as in the rate of growth of goods and services. For as the world's population raises its standard of living, natural resources such as air, water, and minerals will become increasingly scarce. These interrelationships are most important, as our experience with smog and congestion is teaching us.

It is appropriate to analyze the problem of population and production and their interdependence in terms of the concept of optimal population. That concept can clarify and expand some of the statements regarding growth made earlier in this book. On the basis of that analysis and its conclusions, we can then formulate public policies for the future in the United States.

The term "optimal population" usually refers to that population size which maximizes income per person. In general, this relationship hinges on two basic economic concepts—economies or diseconomies of scale (mass production methods) and diminishing social marginal product or its obverse, increasing social marginal cost. The former refers to assembly-line methods of production, which are possible when the market is very large. The latter is the result of expansion of some inputs like labor and machinery, given a limited supply of some other input such as natural resources.[2]

In the long run which we are considering here, the only inputs that remain limited are natural resources (minerals, air, water, land, light, and so on) and management (private and public). In manufacturing and agriculture, as well as in the course of living, wastes must be disposed of. They go into the land, air, or water. Initially, when the country is sparsely populated and not highly urbanized or industrialized, the level of contamination is tolerable. As production increases, however, given the limited supply of natural resources, pollution becomes harmful and must be controlled. More resources must be spent to dispose of the wastes in an innocuous manner. In a complex urban society, these costs increase rapidly, and more than proportionately, as production increases. For example, alternatives to the internal combustion engine must be developed for transportation of people.

Up to a certain point, economies of scale outweigh increasing social marginal costs in their effect on cost per unit of output. Ultimately, however, increasing social marginal cost will more than offset the advantages resulting from economies of scale given the limitation of natural resources.

For our analysis, perhaps a better concept than optimal population is optimal production, that is, that level of production which minimizes average social cost. The reason is that increased production can be the result of a growing population with unchanging per capita income, a population of constant size but with rising per capita income, or an increasing population with increasing per capita income.[3]

In the highly developed countries the benefits of economies of scale are generally prevalent via trade or via the large internal population with high incomes. Given the limited supply of natural resources, however, it appears that there is now a problem of increasing net marginal social cost in the usage of land, air, and water. In other words, more people result in more production and more pollution. To correct this, increasing additional expenditures are needed.

A consequence of this constraint on production is that some of our life-styles may change. One change that seems most probable is that there will be less reliance on individual ownership of durable consumer goods. Indeed, we have created an inefficient system to avoid what might have been a natural evolution to a system which would serve our future needs. For example, we have treated the automobile as an individual convenience, and we have built the whole society around it. Gone are grocery deliveries and other services that the middle class, at least, once enjoyed. We may want to reinstitute some of our grandparents' ways.

It is obvious that even in the United States many people seek and need more income. Given this urge for more and better things, the problem of pollution and the increasing cost of correcting it will continue. If population is stationary, however, the increased production of goods and services and the ensuing pollution are associated solely with higher income per person, and are not the result of more persons. Given a fixed level of pollution, the maximum which is tolerable, income per person could be raised but not indefinitely.

Even if technological change permits the disposal of more waste in a relatively harmless manner, there are still limits, although at a higher level of production. And with a stationary population, this higher level of production can result in higher income per person rather than merely in more production for more people. It appears, therefore, that the present size of population for the United States should not be increased, given the goal of higher income per person and the constraints of pollution and fixed resources.[4]

If the maximum level of pollution is reached and there are no more technological improvements which can reduce that level, then society will have to accept a zero *production* growth rate. Before that point is reached, however, it is preferable to achieve a zero population growth rate. Many presentations have been made in favor of a growing population, so it is well to examine some of them in detail.

The common argument in favor of more people, regardless of income levels, is national power. It is based on the assumption that quantity rather than quality counts in international politics. This argument is contradicted by the fact that a country's influence in the world is now determined by its technological achievements and productive capacity rather than by the mere number of bodies. And economic achievements require physical capital—machinery and factories—as well as a skilled and educated labor force.

The growth and expansion of industry and education are hindered by a rapid growth of the population. High fertility and low mortality result not only in high growth rates but also in a large proportion of young people who are consumers and not producers. The consequence of this is that relatively more is consumed and less is invested, all other things being equal. And this higher proportion of consumption in the aggregate does not necessarily mean higher per capita consumption either.

Suppose that two populations have the same size labor force but one population has more children and therefore more dependents than the other one. Also, assume that the gross national product (GNP) is the same in both economies. If consumption per person is the same in both populations, the one with fewer dependents will consume less of its GNP than the other. As a consequence, there will be more goods and services left over for investment purposes after consumption, if we assume that goods are used either for consumption or investment. Table 6 illustrates this.

It is sometimes argued that more people are needed to initiate economic expansion. In other words, some population growth rate, and not necessarily a high one, is needed to stimulate investment. It is asserted that additional people will motivate the populace to innovate and reorganize their economy so that further economic development is possible. It is a matter of opinion and hard to document, but it seems to us that an argument can just as well be made that innovation stems from other sources, such as education, massive political pressures, or a charismatic leader. Examples of

TABLE 6

Impact on Total Investment of Changing Levels of Dependent Populations

Population	Number in Labor Force	Number of Dependents	Total Population	GNP	Total Consumption at 1,000 units per person	Total Investment
A	1,000	3,000	4,000	5 M*	4 M	1 M
B	1,000	2,000	3,000	5 M	3 M	2 M

*M: million

each of these factors could be given to support one's point of view.

Another argument against a stationary population is that the society will be less prone to change because the older people predominate. The proportion of young people is less in a population that has a low level of fertility. It is also pointed out that in such an economy there will be less expansion of business and government activity. Consequently, chances for promotion and advancement are more limited. Here again, the desire for and achievement of change and growth are not solely functions of the number or proportion of young people. Many other factors are crucial. In such a society, the people may have to change some of their attitudes and use new approaches for the maintenance of an active and vital population rather than rely on a high birth rate to provide a stimulus.

With zero population growth, the government can concentrate all its additional expenditures on improving the quality of its constituents' lives. No additional funds are needed for more extensive facilities for more people. Rather, more intensive programs can be initiated such as those in the field of mental health facilities for children and adults, day-care centers, and the like, and in the elimination of slums and poverty pockets. All these would help meet the need for improved quality of the lives of many people—something desirable even in the richest of countries. Every country has problems, and these are more readily soluble when the number of potential problem makers is not increasing.

Another favorable aspect of zero population growth is that if population is not increasing, wages will tend to rise and the rent for land will rise less rapidly. A consequence of this should be some redistribution of income from the higher to the lower income groups. This follows from the concentration of land ownership among relatively few people, whereas the majority of people work for wages. The problems of bulging age groups would also be avoided.

In a stationary population there are proportionately more elderly people. The problems of the aging are known to everyone, and one of the major causes of the difficulties suffered by old people is the failure of society to make them feel useful. Given an expanding economy and a stationary population, it may well be that society can continue to use the elderly in the labor force on a part-time basis or in a variety of minor capacities as a result of the relative scarcity of labor.

In summary, the arguments in favor of a growing population do not seem as convincing as the arguments in favor of a stationary population. This is all the more persuasive given the urge for more and better things, i.e. higher incomes, and the propensity for urban living. Very few people are satisfied with their current level of income. Yet the constraint of pollution and fixed resources, and the increasing cost of overcoming it, indicate that production cannot increase forever.

One way to curb growth in production is to achieve a zero population growth rate. We are now approaching that on a voluntary basis in the United States. Hopefully, other nations will respond appropriately. After the attainment of that goal, however, we must consider a diminution in the growth rate of production. Given the poverty of most of the world, their GNP growth rates should continue high for many years. In the United States it is a different story. For the reasons given in the earlier analysis, we think our growth rates should gradually diminish.

Even though a stationary population permits a higher standard of living with relatively less pollution, ultimately the restraints imposed by natural resources will force a slowing down of economic expansion. This will be a gradual process and, for that reason, it can be managed. As long as 20 percent of our population is living on the margin of our economy, however, we cannot accept an upper limit on the GNP and expect the present distribution of income to prevail in perpetuity. What we must work on is a redistribution of income or at least a redistribution of service so that the gap between the very rich and the very poor is reduced. Then we would have a more equitable economic system.

How can these two proposed goals of a reduction in GNP growth rates and a redistribution of income be achieved? Are they not incompatible? The following section examines these two questions.

Declining Growth of GNP

As the growth rate of the economy declines, net investment as a percent of GNP must decline by definition. Net investment is defined as the change in buildings, equipment, and inventory after replacement of capital that has depreciated. The greater the proportion of GNP that is invested, the greater the relative growth of the GNP. If the rate of growth is to decline, therefore, so must the proportion of GNP that is invested.

For an economy to have an equilibrium level of income at a full employment level (or any other level), planned investment must equal planned savings at that income level. Therefore, the maintenance of a full employment level of income requires that as investment declines so must savings. Savings are defined as those goods and services which are not consumed. And what is not consumed by private individuals or the government must be invested (excluding the export sector which is about 4 percent of our total production).

The question, then, is one of raising the proportion of GNP that is consumed by the public and private sectors in order to reduce the surplus or savings available for investment. It is this difference between total GNP and private consumption plus government consumption, which we defined as savings, that must be invested in order to maintain income at that level. How, then, can we raise the proportion consumed?

People with higher incomes inevitably save a larger proportion of their income than do the poor. They have more life insurance, stocks, savings accounts, and pension plans. The poor are forced to consume all their income in order merely to survive. So, a transfer of income from the more affluent to the less affluent would raise the proportion of GNP that is consumed.

One way to do this is to guarantee a minimum income to everyone. This could be financed by altering the tax burden. Such an adjustment might involve nothing more than plugging the loopholes used by those with incomes over $25,000 a year, as well as all the loopholes exploited by corporations. Their tax shelters and their "welfare" programs would have to go. The money gained could easily finance a guaranteed annual income of $5,000–$6,000 per family. And those recipients could be counted on to spend it all, as they now live in slums and eat inadequately.

Another approach would be to impose a limit on personal income in the sense that everything earned over, say, $100,000 was taxed at the rate of 100 percent. What this would accomplish would be a redistribution of the highly paid jobs. People may as well quit working after they have earned the $100,000 limit for the year. For the remainder of the year, say four or five months, other people who are also competent but have not yet earned the $100,000 maximum could fill these highly paid positions. In this way, the work would get done and the highly paid jobs would be spread out over more and more people. The only prerequisite here is that there must be an ample supply of trained people for skilled and professional positions in such fields as medicine and the sciences, and on managerial levels. If more openings were available for training such people, there would be no dearth of interested and qualified applicants.

This same goal might also be achieved by reducing the hours of work from eight hours a day to perhaps five or six hours a day. As income rises, some of the gains in productivity can easily be taken in the form of more leisure rather than more pay. This approach could apply to all income groups, not just those at the top.

A general reduction in the working week would be most beneficial to all if the increased leisure meant more time for human development. As we approach zero population growth via the two-child family, and as more women enter the labor force, shorter working hours will leave more time for both parents to spend with their children. In fact, when the economy is no longer expanding so rapidly, men may have to share some of their hours with the increased number of women demanding jobs.

Savings can also be reduced on the corporate level. Corporations save billions of dollars a year in the form of retained earnings. This is money withheld from profits and not distributed to stockholders. About 50 percent of corporate investment is financed by funds generated internally. The amount of retained earnings is dependent on profits. A reduction in profits, therefore, would result in a reduction in savings. Our proposal, PSP, to stabilize and limit profits would also stabilize and limit corporate savings. This in turn means that the profit limit could be adjusted so as to reduce corporate savings should it be necessary.

When there is less emphasis on the need for high profits, some of the incentives for ruinous competition and ruthless exploitation will be lessened. More business energy and time can be devoted to service, to taking pride in one's product, and to the humanizing of the working environment. In this way we can overcome our slavery to material things and, instead, use them for our further human development.

This new approach to business activity will ultimately have to be extended to the rest of the world. Once PSP is adopted in the United States, we propose that the United States negotiate with other leading countries an agreement on the international application of PSP. In this way, an international corporation could not play one country against another when choosing a corporate "home." A large part of the tax revenue collected from international companies whose profits exceeded the maximum could then be pumped into the international lending agencies, such as the

World Bank (IBRD) and regional development banks. This would free the domestic taxpayers in developed countries from some of the costs of necessary world development. Consequently, some of the money taken out of developing countries by foreign corporations would automatically go back to those countries. This would even help the foreign corporations, as the developing countries would be able to provide more services in the form of roads, power, schools, and the like. Thus the principle of public interest based on equitability would be extended to the world economy. As a result, private foreign capital would be more welcome than it now is.

If this idea becomes feasible internationally,[5] investments of private corporations might have a domestic side, subject to the PSP rules, and a foreign side, subject in part to international rules based on something like PSP. No corporation would then be punished domestically for its foreign successes. Nor would the inefficiency of domestic branches of the corporation be hidden from the stockholders' view as is now the case. This would help to guarantee that the owners of American capital remained interested in promoting efficiency and productivity in the domestic branches of their corporations, rather than always seeking greener fields abroad. The lesson of British experience should not be lost on us. It was the British corporation, installing all the latest British equipment abroad and none within the home country, that tipped the balance of productivity away from Britain and toward her commonwealth competitors.

Conclusion

Thus, stability without growth is possible for the United States at a high level of personal income and well-being, even in a world where rapid economic growth of developing areas is still the goal. What John Stuart Mill saw in the mid-nineteenth century as a possibility—"the stationary state"—is not to be feared but planned for. As we have explained, overcoming some of the constraints on growth is beyond even modern technical capacities. Admitting this, we can construct alternatives to our obsession with more people and more things.

There is an even greater challenge. In a world where some believe that stability, economic planning, and the solution to economic problems are possible only through the surrender of power to a few, we can continue to pursue our value of dispersed power among the many. One manifestation of this would be adoption of PSP. We can reach new levels of performance, in part through education and understanding, with a modicum of controls on greed. For those who assume that evolution of values is not possible, that man is but an impossibly violent and aggressive animal, the future is not nearly so hopeful.

Our optimism is based on the belief that man has over the few centuries of his written history demonstrated a capacity for mental growth and adaptation which would make the spaceship earth livable for many more centuries. Our suggestions do not exhaust the possibilities for these adjustments. We have not hinged our hopes, for instance, on the possibility of a total cessation of war, though the cessation of war seems to us a reasonable and even attainable goal. We have not argued that the disappearance of political boundaries is necessary to solve what are really global problems. We are arguing rather that the United States, in solving some of its problems in a constructive way via PSP, can join other enlightened countries in making the world economy a more stable and equitable one. It is clearly in our public interest and within our capacity to do so.

Appendix of Related Readings

CHAPTER 1

READING 1

*Excerpts from a Debate between Milton Friedman and Robert Solow upon the Occasion of the Johnson Wage-Price Guideposts**

Friedman: What Price Guideposts?

Since the time of Diocletian, and very probably long before, the sovereign has repeatedly responded to generally rising prices in precisely the same way: by berating the "profiteers," calling on private persons to show social responsibility by holding down the prices at which they sell their products or their services, and trying, through legal prohibitions or other devices, to prevent individual prices from rising. The result of such measures has always been the same: complete failure. Inflation has been stopped when and only when the quantity of money has been kept from rising too fast, and that cure has been effective whether or not the other measures were taken.

Entirely aside from their strictly economic effects, guidelines threaten the consensus of shared values that is the moral basis of a free society. Compliance with them is urged in the name of social responsibility; yet, those who comply hurt both themselves and the community. Morally questionable behavior—the evading of requests from the highest officials, let alone the violation of legally imposed price and wage controls—is both privately and socially beneficial. That way lies disrespect for the

---

*George P. Shultz and Robert Z. Aliber, eds., *Guidelines, Informal Controls, and the Market Place: Policy Choices in a Full Employment Economy* (Chicago: University of Chicago Press, 1966), pp. 17–19, 41, 43, 57, 62–63. ©1966 by The University of Chicago.

law on the part of the public and pressure to use extralegal powers on the part of officials. The price of guideposts is far too high for the return, which, at most, is the appearance of doing something about a real problem.

Solow: The Case Against the Case Against the Guideposts

I choose this defensive-sounding title because it points to an important truth. The wage-price guideposts, to the extent that they can be said to constitute a policy, are not the sort of policy you would invent if you were inventing policies from scratch. They are the type of policy you back into as you search for ways to protect an imperfect economy from the worst consequences of its imperfect behavior. For this reason, it seems to me that the best way to start an evaluation of the wage-price guideposts is with recognition of the dilemma to which they are a response.

* * *

The experience of the years 1958–64 certainly indicates that the economy can be run with quite a lot of slack, but not a catastrophic amount, so that the price level will more or less police itself. That is a possible policy. But it is not a costless policy. In the first place, one of the necessary concomitants of this policy is a pretty substantial unemployment rate. Since the incidence of unemployment is typically uneven, and the unevenness has no claim to equity, common decency requires that this policy be accompanied by a major reform and improvement of the unemployment compensation system, and possibly of other transfer payment systems as well. This is a budgetary cost, but not a real burden on the economy as a whole. In the second place, however, the maintenance of slack does represent a real burden to the economy as a whole in the form of unproduced output. It is not easy to make any estimate of that cost. The usual rule of thumb is that one-half point on the unemployment rate corresponds to something between 1 and 2 per cent of real GNP. In that case, the amount of relief from inflation that could be had by keeping the unemployment rate one-half point higher than otherwise desirable would have an annual cost of about $10 billion at 1965 prices and GNP.

Solow: (on the issue of market power)

[One] approach is to recognize that the threat of premature inflation reflects significant departures from perfect competition in labor and product markets: The appropriate remedy is to create or restore competition by breaking up all concentrations of market power, whether in the hands of trade unions or large firms, and by eliminating all or most legal protections against domestic and foreign competition.

[This approach] caters to the economist's prejudice in favor of the mechanism of the competitive market. If there is a case against the case against the guideposts, part of it has to be that [this] obvious remedy is more than a little unrealistic.

Friedman: Response to the Market Power Argument

I want to go to the logic underlying the guideposts. This logic is that there is market power lying around, and that, when times are reasonably good, market power is likely to be exercised in ways that contribute to premature inflation. One thing that always impresses me about this argument is how briefly it is alluded to *when* it is alluded to—at all. In paper after paper in the discussion of guideposts and cost-push inflation, or of market power as a source of inflation, you discover that there is but a sentence or two, and then the author goes on to other things. The reason for that is very clear—the logic of the analysis is wrong. Insofar as market power has anything to do with possible inflation, what is important is not the *level* of market power, but whether market power is *growing* or not. If there is an existing state of monopolies all over the lot, but the degree of monopoly has not been increasing, this monopoly power *will not* and *cannot be* a source of pressure for inflation.

Solow: Comment

Milton and I seem to be talking about wholly different things; certainly, we are talking about them in a wholly different manner. My own attitude to the guideposts is not diametrically opposed to his: when I showed what I had written to my wife, she muttered something to the effect that if those guideposts have you for a friend, they have no need for an enemy.

I do tend to be a lot more tentative about most things than Milton. I think that part of the difference between our papers and our views is a matter of temperament. I have to admit that Milton reminds me a little of what Lord Melbourne is supposed to have said about Macaulay; namely, "I wish I was as cocksure about anything as Macaulay is about everything."

Another difference between Milton and myself is that everything reminds Milton of the money supply. Well, everything reminds me of sex, but I keep it out of the paper.

READING 2

The Issue of Whether Monopoly is an Economic Problem and the Division among Economists on What To Do about the Current Structure of American Industry

Concentration of economic power has long been an issue in systematic economic thought. As early as 1838, Cournot proposed the model of monopoly. Marx relied on what he considered to be a natural tendency, since he observed that capitalists seemed bent on concentration of economic power. Yet, curiously, it was after Marx, rather than before, that the strength of the competitive model and the associated policy of laissez-faire reached its height.

In the United States, whether upon the advice of economists or not, the legislative bodies at both the state and federal levels took a dim view of market power. Market power is the ability of a few dominant companies to influence prices in a particular industry. Hence, since 1890, we have developed in this country a history of trying to cope with trusts, interlocking directorates, and other forms of monopoly power.

Here, as in other questions involving policy decisions, economists have disagreed among themselves regarding both the facts of the case and the remedy, if any, which is needed. These divisions have followed roughly the pattern of interventionism and noninterventionism.

The factual issue is an argument over whether bigness, as such, is a measure of significant market power. If it is, then the vast internal growth and the numerous mergers that have occurred since 1939 have made supergiants of what were corporate giants before. Charles H. Hession and Hyman Sardy conclude:

> Certainly, if bigness is the criterion, our corporations increasingly deserved this characterization. Consider that in 1947, there were 113 corporations with assets of $100 million or more; by 1962, their number had more than tripled to 370. General Motors, the nation's biggest industrial company, is bigger today than most nations. In fact, its net operating revenue of $11 billion in 1965 exceeded the 1964 gross national product of all but nine nations of the free world. The increase that took place in overall concentration as accounted for by our largest manufacturing corporations between 1947 and 1962 is shown in Chart 4.[1]

But many economists deny that bigness or overall concentration of assets or ownership is the significant measure. They study concentration by industry, for example, and show, as we see in table 7, that since 1945 there has been little *increase* in industrial concentration in the sense of *concentration of the percentage of sales accounted for by the 4 largest firms* in 400 industries.[2]

Hession and Sardy are concerned about the apparent conflict between increasing overall concentration and stable or even declining "average" concentration by industry. The conflict is resolved if one examines the form which mergers have taken in product extension. The conglomerate, which they call the "latest corporate mutation in the continuing 'organizational revolution' in the United States," extends power to diverse industries. Testimony in Senate hearings shows that this trend toward diversification became pronounced after 1954; see table 8.[3]

The disagreement on the relevant facts does not end the argument. Philip A. Klein reviews the policy implications of trying

CHART 4

Share of Value Added by Manufacture Accounted for by 200 Largest Manufacturing Companies, 1947-1962

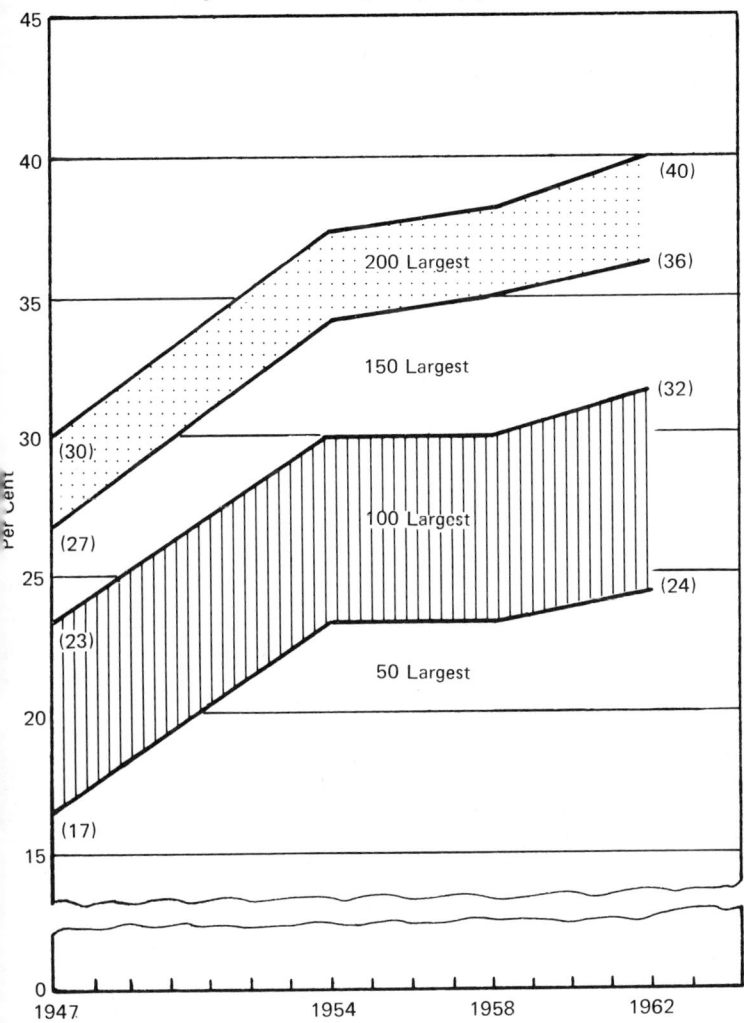

Source: U.S. Bureau of the Census

TABLE 7

Percentage of Sales Accounted for by the 4 Largest Firms in 400 Industries

	Industries Having a Ratio of 75% or More	Industries Having a Ratio of 50% or More
1947	10.0%	23.7%
1954	9.3	24.7
1958	7.9	22.6

TABLE 8

Distribution of Large Manufacturing and Mining Acquisitions, by Type and by Period of Acquisition

	1948-53 Number	1948-53 Percent	1954-59 Number	1954-59 Percent	1960-64 Number	1960-64 Perce
Horizontal	18	31.0%	78	24.8%	42	12.
Vertical	6	10.3	43	13.7	59	17
Conglomerate						
Market extension	4	6.9	20	6.4	24	6.
Product extension	27	46.6	145	46.2	184	52
Other	3	5.2	28	8.9	39	11.
Totals	58	100.0%	314	100.0%	348	100.

to distinguish between market *size,* market *conduct,* and market *behavior.* The antitrust record is checkered in regard to size. In the Standard Oil case of 1911, the Supreme Court declared that "mere size is no offense," but in the 1940s, doctrines began to develop that mere size, if it carries with it sufficient market control, can be illegal.[4]

Market conduct rules have sometimes been more vigorously enforced against the behavior of several firms than against very large firms that can accomplish the same objectives by virtue of market power:

Thus a large firm controlling 50 percent of the output of some product clearly sells at a price which it sets. More often than not the firm has no problem with the Department of Justice. However if two firms, each of which controls 25 percent of the product market, make an *agreement* to sell at the same price, all hell breaks loose. The *result* is virtually the same, but the antitrust laws have obviously been interpreted much more strictly against restrictive practices reached through collusion than through mere size.[5]

Klein believes that a fourth standard is emerging both in the United States and abroad, that is:

. . . a tendency to pay increasing attention to the market *performance* of various industries and to judge the success or failure of public policy by pragmatic results rather than by either structure or behavior criteria alone. Thus one can ask: Are most plants of reasonably efficient size? Is there an acceptable level of excess capacity in plants? *Are price-cost margins such that profits can be considered "normal" rather than "monopolistic" or "excessive"?* Are there pressures producing unremitting efforts to improve both production processes and products, with subsequent benefits passed on at least partly to consumers? Are advertising and other selling efforts considered "reasonable"? [Italics added.][6]

NONINTERVENTIONIST VIEWS:

Friedman's views on monopoly rest on his consideration of the facts as well as his ideological predisposition: "The most important fact about enterprise monopoly is its relative unimportance from the point of view of the economy as a whole." He adds that one source of the "mistaken" impression that monopoly is important is the "tendency to confuse absolute and relative size," and another that "monopoly is more newsworthy . . . than competition."[7]

His remedy is also related to what he considers "causes" of monopoly: "Probably the most important source of monopoly power has been government assistance, direct and indirect."[8] It follows that the "first and most urgent necessity in the area of government policy is the elimination of those measures which directly support monopoly, whether enterprise monopoly or labor monopoly."

Included in this solution would be extensive reform of the tax laws, including the abolition of the corporate tax:

> The corporate tax should be abolished. Whether this is done or not, corporations should be required to attribute to individual stockholders earnings which are not paid out as dividends. That is, when the corporation sends out a dividend check, it should also send a statement saying, "In addition to this dividend of _____ cents per share, your corporation also earned _____ cents per share which was reinvested."

He concludes that such a law would do much "to invigorate capital markets, to stimulate enterprise, and to promote effective competition." In addition, he would scale down the higher rates of income tax and eliminate "the avoidance devices that have been incorporated in the law." As for antitrust, he would "treat both enterprise and unions alike."[9]

Galbraith, while an interventionist on some issues, appeared before the Senate Subcommittee of the Select Committee on Small Business and urged that "the trend to great size and associated control was immutable, given our desire for economic development," that "the present anti-trust efforts to deal with size and market power were a charade" and, finally, that "the anti-trust laws legitimize the real exercise of market power on the part of the large firms by a rather diligent harassment of those who have less of it."

But unlike Friedman, who discounts bigness, Galbraith positively defends it from some sort of "preventive" action to try to keep smaller firms from getting larger: "I am content to argue that we have big business, and that the antitrust laws notwithstanding we will continue to have it, and that they give an impression of alternative possibilities that do not exist."[10] Walter Adams counters with the accusation that Galbraith's new industrial state is "a blueprint for technocracy, private socialism, and the corporate state." Adams agrees that corporate giantism dominates American industry, but "Galbraith fails to prove that this dominance is the inevitable response to technological imperatives and hence beyond our control." Thus, says Adams, there are more attractive public policy alternatives than Galbraith suggests.[11] Adams believes that one can favor technological bigness, for example, while opposing administrative bigness, and without inconsistency. Like Friedman, Adams thinks the government should look to its own policies for a solution: "A competitive society is the product not simply of negative enforcement of the antitrust laws; it is the product of a total integrated approach on all levels of government—legislative, administrative, and regulatory. An integrated national policy of promoting competition—and this means more than mere enforcement of the antitrust laws—is not only feasible but desirable."[12]

So the curious tangle. Galbraith thinks that planning has replaced or must replace the market so far as the industrial system is concerned. Adams cites the electric power industry as an example of his belief that "monopoloid planning is done in the interest of monopoly power; seldom, if ever, is society the beneficiary." Galbraith thus thinks that the mature corporation must be free to plan, free that is, of market and antitrust constraints. Adams claims that the force of competition is "superior to industrial planning—by private monopolist, the benevolent or authoritarian bureaucrat."

One must conclude as one began, that economists agree neither on the facts nor the programs when questions are raised about the current structure of American industry.

CHAPTER 2

READING 3

Issues in Freedom and Control: What Economists Think about Planning and Public Interest

The concept of public interest raises the issue of "common welfare," which has been debated by economists throughout their history. If we limit our concern to the post-World War II debate, the issues involve the degree, if any, of planning that is compatible with freedom both of the market and of the individual.

Friedrich A. Hayek's celebrated *The Road to Serfdom* condemned not only socialism as a system of state planning but virtually any form of state economic planning. In effect he argues that the "Rule of Law," which implies limits to the scope of legislation, must exclude "legislation either directly aimed at particular people or at enabling anybody to use the coercive power of the state for the purpose of such discrimination."[1]

Two examples Hayek uses are of special contemporary significance. He admires the statement from H. G. Wells's "Declaration of the Rights of Man" that every man "shall have the right to buy and sell without any discriminatory restrictions anything which may be lawfully bought and sold." But Hayek accuses Wells of making the whole provision nugatory by adding that this only applies to buying and selling "in such quantities and with such reservations as are compatible with the *common welfare.*"

Hayek's second example, which, again, refers to a quotation from Wells, is relevant to current debates on equal opportunity:

> Or, to take another basic clause, the declaration states that every man "may engage in any lawful occupation" and that "he is entitled to paid employment and to a free choice whenever there is any variety of employment open to him." *It is not stated, however, who is to decide whether a particular employment is "open" to a particular person,* and the added provision that "he may suggest employment for himself and have his claim publicly considered, accepted or dismissed," shows that Mr. Wells is thinking in terms of an authority which decides whether a man is "entitled" to a particular position—which certainly means the opposite of free choice of occupation. [Italics added.][2]

J. M. Clark countered Hayek's approach in *Alternative to Serfdom*, calling for a "balanced society":

> We have been living in a world of conflict between false

absolutes—the absolute community or the absolute state, and the absolute individual. Our liberal civilization has been built on the myth of the absolute individual, whom the state and the community exist to serve; the community being an arithmetic sum of such individuals, and the state their agent, serving them best by giving them maximum liberty to serve themselves. Over against this theory, and taking advantage of its excesses and shortcomings, has arisen the doctrine of the totalitarian state, under which the individual exists to serve the community, of which the state is the embodiment; and the state's power embraces everything in life. In practice this means power that is not only unlimited, but irresponsible.

Our abhorrence of this doctrine should not lead us to support its extreme opposite, which leads to the apotheosis of irresponsible private self-interest. For neither of these one-sided theories is the truth about man and society, and neither is a sound basis for building a social constitution.[3]

Galbraith introduced the principle of countervailing power into the argument, essentially an evolutionary argument that one thing (big business) calls for another (big labor) and the existence of these privately powerful groups requires a third (big government).[4] But more recently, he has qualified this view, as we shall see.

Meanwhile, Hayek's arguments, which rose mainly through the Austrian anticommunist school of economic thought, found a philosophical home in the United States in what is called the Chicago school of economics, of which Friedman is alleged to be a member. Thus Friedman, in his *Capitalism and Freedom,* gave homage to Hayek, but moved his arguments to newer conditions. Friedman argued that we now have several decades of experience with governmental intervention:

. . . it is clear that the difference between the actual operation of the market and its ideal operation—great though it undoubtedly is—is nothing compared to the difference between the actual effects of government intervention and their intended effects.[5]

He does not think it an accident that government reforms have failed, as he sees it, to achieve their objectives. The failure was guaranteed, he feels, any time government tried to substitute the "values of outsiders for the values of participants." He cites, in addition to the "evil men in the Kremlin," the "internal threat coming from men of good intentions and good will who wish to reform us."[6] He concludes that while national-defense buying by the federal government is inevitable, we should not extend governmental intervention to the undertaking of "ever new governmental programs—from medical care for the aged to lunar exploration." Hence one detects that Friedman's definitions of proper subjects for governmental intervention are even narrower than those of Adam Smith. For Smith had conceded a need for government wherever competition failed. Clearly, Friedman's view is one that leaves little scope for public interest outside of military posturing.

Galbraith takes a rather different view of the experience of recent years. He sees that "the industrial system, in fact, is inextricably associated with the state." In *The New Industrial State*, Galbraith theorized that the historical fears that the entrepreneurial enterprise would dominate the state were replaced by fears that the state would dominate the entrepreneurial enterprise. But meanwhile this enterprise was being replaced by the evolving mature corporation or "technostructure." What finally emerges is a symbiotic relationship between the goals of the mature corporation and those of the state:

> The state is strongly concerned with the stability of the economy. And with its expansion or growth. And with education. And with technical and scientific advance. And, most notably, with the national defense. These are *the* national goals. . . . All have their counterpart in the needs and goals of the technostructure.[7]

Thus Galbraith deemphasizes the role of economic conflict. Yet, unlike Friedman, he leaves room for evolving social goals, and thus for new and more encompassing definitions of public interest.

Joan Robinson seconds Galbraith's notions, concluding in her *Freedom and Necessity* that

> the economists of the laissez-faire school purported to abolish the moral problem by showing that the pursuit of self-interest by each individual rebounds to the benefit of all. The task of the generation now in rebellion is to reassert the authority of morality over technology; the business of social scientists is to help them to see both how necessary and how difficult that task is going to be.[8]

This debate on public interest by economists is carried out on two levels: (1) the philosophical or ideological, which investigates the meaning of planning to the individual and the state; (2) the empirical, which allows economists to disagree on the meaning of historical events.

Events themselves cannot wait for the debaters. Evolutionary economists such as Philip A. Klein note that present economic systems "defy the neat characterizations to which economists of another day might have been lured in their attempts to distinguish capitalism, socialism, communism, and other isms." He asks us to remember Clark once more, "who argued that modern market-oriented economies had no choice but to utilize the market, the state and the other organized groups and develop a technique for coordinating them all."[9]

READING 4

The Problem of Inflation as Arthur Burns Sees It

Fiscal and monetary policies are not sufficient for dealing with inflation. Recent history supports such an assertion, and Arthur F. Burns, chairman of the board of governors of the Federal Reserve system, also appears to agree. We would like to quote excerpts from an excellent address he made at the ASSA (Allied Social Science Associations) Convention in Toronto, Canada, on December 29, 1972. It is entitled "The Problem of Inflation."

Burns's diagnosis of the ills in our economic system that result in inflation is very perceptive. When it comes to a positive program for coping with them, however, he relies on more of the same: improved monetary and fiscal policy; reduction in the growth of federal spending; continuation of wage and price controls for a while longer; and structural reforms that lead to more competition.

We feel that if he had followed his analysis to its logical conclusion, he would have recommended some type of plan whereby price increases would be tied to profits. Economists are generally reluctant, however, to interfere with profits or even to discuss their restraint. Because of social conditioning, perhaps, it is as painful for us, the capitalists, to consider restriction of profits as it is for communists to consider their encouragement as an incentive for good work.

The Problem of Inflation

The current inflationary problem has no close parallel in economic history. In the past, inflation in the United States was associated with military outlays during wars or with investment booms in peacetime. Once these episodes passed, the price level typically declined, and many years often elapsed before prices returned to their previous peak. In the economic environment of earlier times, business and consumer decisions were therefore influenced far more by expectations concerning short-term movements in prices than by their long-term trend.

Over the past century, a rather different pattern of wage and price behavior has emerged. Prices of many individual commodities still demonstrate a capability of declining when demand weakens. The average level of prices, however, hardly ever declines. Wage rates have become still more inflexible. Wage reductions are nowadays rare even in ailing businesses, and the average level of wages seems to rise inexorably across the industrial range.

The hard fact is that market forces no longer can be counted on to check the upward course of wages and prices even when the aggregate demand for goods and services declines in the course of a business recession. During the recession of 1970 and the weak recovery of early 1971, the pace of wage increases did not at all abate as unemployment rose, and there was only fragmentary evidence of a slowing in price increases. The rate of inflation was almost as high in the first half of 1971, when unemployment averaged 6 per cent of the labor force, as it was in 1969, when the unemployment rate averaged 3 and a half per cent.

The implications of these facts are not yet fully perceived. Cost-push inflation, while a comparatively new phenomenon on the American scene, has been altering the economic environment in fundamental ways. For when prices are pulled up by expanding demands in times of prosperity, and are also pushed up by rising costs during slack periods, decisions of the economic community are apt to be dominated by expectations of inflation.

* * *

In fact, the outcome of our struggle with inflation is likely to have world-wide repercussions. If we continue to make progress in solving the inflation problem, our success will bring new hope to other countries of the Western world where inflationary trends stem in large measure from the same sources as ours.

Almost the entire world is at present suffering from inflation, and in many countries—for example, Canada, France, the United Kingdom, West Germany, and the Netherlands—the pace of inflation is more serious than in the United States.

* * *

These countries have discovered, as we in the United States have, that wage rates and prices no longer respond as they once did to the play of market forces.

As I have already noted, a major cause of the inflationary bias in modern industrialized nations is their relative success in maintaining prosperity. Governments, moreover, have taken numerous steps to relieve burdens of economic dislocation. In the United States for example, the unemployment insurance system has been greatly strengthened since the end of World War II: compensation payments have increased, their duration has lengthened, and their coverage has been extended materially, thus easing the burdens of retirement or job loss for older workers, and welfare programs have proliferated.

Protection from the hardships of economic displacement has been extended by government to business firms as well. The rigors of competitive enterprise are nowadays blunted by import quotas, tariffs, price maintenance laws, and other forms of governmental regulation; subsidy programs sustain the incomes of farmers; small businesses and home builders are provided special credit facilities and other assistance; and even large firms of national reputation look to the Federal Government for sustenance in times of trouble.

Thus, in today's economic environment, workers who become unemployed can normally look forward to being rehired soon in the same line of activity, if not by the same firm. The unemployment benefits to which they are entitled blunt their incentive to seek work in an alternative line or to accept a job at a lower wage. Similarly, business firms caught with rising inventories when sales turn down are less likely to cut prices to clear the shelves—as they once did. Experience has taught them that, in all probability, demand will turn up again shortly, and that stocks of materials and finished goods—once depleted—nearly always have to be replaced at higher cost.

Institutional features of our labor and product markets reinforce these wage and price tendencies. Excessive wage increases tend to spread faster and more widely than they used to, partly because workmen have become more sensitive to wage developments elsewhere, partly also because employers have found—or come to believe—that a stable work force can best be maintained in a prosperous economy by emulating wage settlements in unionized industries. In not a few of our businesses, price competition has given way to rivalry through advertising, entertaining customers, and other forms of salesmanship. Trade unions at times place higher priority on the size of wage increases than on the employment of their members, and their strength at the bargaining table has certainly increased. The spread in recent years of trade unions to the public sector has occasioned some illegal strikes which ended with the union demands, however extreme, being largely met. The apparent helplessness of governments to deal with the problem has encouraged other trade unions to exercise their latent power more boldly. And their ability to impose long and costly strikes has been enhanced by the stronger financial position of American families, besides the unemployment compensation, food stamps, and other welfare benefits that are not infrequently available to strikers.

In view of these conditions, general price stability would be difficult to achieve even if economic stabilization policies could prevent altogether the emergence of excess aggregate demand. But neither the United States nor any other Western nation has come close to that degree of precision. In fact, excess aggregate demand has become rather commonplace. In country after country, stabilization efforts have been thwarted by governmental budgets that got out of control, and central banks have often felt compelled to finance huge budgetary deficits by credit creation.

* * *

The only responsible course open to us, I believe, is to fight inflation tenaciously and with all the weapons at our command. Let me note, however, that there is no way to turn back the clock and restore the environment of a bygone era. We can no longer cope with inflation by letting recessions run their course; or by accepting a higher average level of unemployment; or by neglecting programs whose aim is to halt the decay of our central cities, or to provide better medical care for the aged, or to create larger opportunities for the poor.

A modern democracy cannot ignore the legitimate aspirations of its citizens, and there is no need to do so. The rising aspirations of our people are consistent with general price stability if we only have the will and the good sense to pursue an appropriate public policy. Our needs are, first, to restore order in the Federal budget and strengthen the stabilizing role of fiscal policy; second, to pursue monetary policies that are consistent with orderly economic expansion and return to a stable price level; third, to continue for a while longer effective controls over many, but by no means all, wage bargains and prices; and fourth, to reduce or remove existing impediments to a more competitive determination of wages and prices.

READING 5

Shortages, PSP, and Price Control

Some people may charge that PSP could cause shortages. They might compare our proposed restrictions on price increases to those imposed by the government on retail beef prices during the summer of 1973, when some meat-packers operated at a loss because retail beef prices were temporarily frozen whereas the price of cattle was allowed to rise. Profits were reduced as a consequence, and many meat-packers closed their plants. It might be argued that something similar could happen if PSP were adopted.

Under PSP, a large company in a noncompetitive industry can earn up to a 10 percent return on its assets after normal taxes. This is based on a five-year moving average. The company cannot raise its prices unless that return drops to 6 percent. This is the profit range we have proposed for an industry dominated by a few large firms.

It is conceivable that profits could drop to zero or a negative value for a short period in a particular industry. This could be the result of some unforeseen and unique event such as a natural disaster or the sale of massive quantities of goods to a foreign government, such as our wheat sales to the Soviet Union. Such huge transactions inevitably dislocate the smooth operation of any market.

Let us assume that something of this type occurs. The consequence for a hypothetical company, company A, is that the price of inputs or raw materials has risen sharply. Given our proposal, prices charged by company A for its final products cannot be raised until the previous five-year average of profit drops to a rate below 6 percent. Assume that company A has been earning the maximum allowable return of 10 percent for the past five years. In addition, assume that the profit rate after the dislocation described above drops to zero or a negative rate and will continue at that level unless prices can be raised. Given these hypothetical conditions, company A cannot raise its prices until after about two years of zero or negative profit rates. This is the case because only then would the average drop below 6 percent. If we assume that the profit rates had been 10 percent for each of three years, 0 percent for the fourth year, and −½ percent for the fifth year, then the average would be less than 6 percent.[1]

Company A might choose to close its factory rather than suffer losses if the price of the raw material was expected to drop in the future. Such would be the case with our example of agricultural products after the supply increases and the excessive demand is reduced following the completion of the sale to a foreign government. In the case of beef, however, the meat-packers were awaiting the removal of the price ceiling on beef so that prices and profits could be increased.

If company A and others like it discontinued their operations, then such shortages would develop. Prices would not rise but neither could goods be purchased in the stores.

If something of this nature occurs and PSP is in effect, then the following corrective action can be taken. The price constraint aspect imposed by PSP on concentrated industries can be waived so that the prices can be raised. It is to be remembered that we are discussing a concentrated industry where a few companies dominate and not a competitive one where market forces determine prices because of the existence of many sellers.

Given such price increases, company A's profits could rise from zero to a positive level. The tax aspect of PSP, however, would not be waived. All profits after normal taxes in excess of a 10 percent return would be taxed at the rate of 100 percent. Companies, therefore, would have no incentive to raise prices above the level sufficient to earn that maximum 10 percent return. When conditions return to a normal state, the constraint on price increases making them contingent on profit rates could be reimposed.

One further adaptation could also be utilized. We have proposed that smaller companies be allowed a higher maximum profit level, such as 14 percent, before the excess is taxed. If shortages developed in a particular industry dominated by a few companies, then new and independent companies could be induced to operate in this field by offering them the 14 percent maximum return, or more, if necessary, so long as the new companies remained small relative to the total market. In this way, an industry could be made more competitive by encouraging the entry of new business. When an industry becomes more competitive because of the presence of more firms, profit rates tend to decline because many companies are now competing with one another for customers.

When an industry is competitive, PSP imposes no price constraints but lets prices rise and fall according to market forces. But PSP does tax profits in excess of some predetermined percentage even in competitive industries.

In effect, what we have proposed here to correct any shortages that might arise is to treat an industry with concentrated power in the same way as a competitive industry. Then we encourage the entry of more companies into that industry to make it more competitive. If such efforts succeed, then that industry can continue to be treated as a competitive industry. If not, when the crisis has passed that industry can be regulated once again in the same way as one with concentrated market power is regulated.

This example illustrates the flexibility of PSP and its ability to deal with a variety of market circumstances.

CHAPTER 3

READING 6

Excerpts from the British White Paper of March 1973, "The Counter-Inflation Programme," "The Operation of Stage Two," Presented to Parliament by the Chancellor of the Exchequer by Command of Her Majesty, March 1973.

PRICES AND PROFIT MARGINS

61. Prices should be determined so as to secure that net profit margins, as defined in paragraph 62, do not exceed the average level of the best two of the last five years of account of the enterprise concerned ending not later than 30 April 1973 (the "reference level").

62. "Net profit margin" means the margin of net profit expressed as a percentage of sales or turnover. "Net profit" means the net profit, determined in accordance with generally accepted accounting principles consistently applied by the enterprise concerned, which arises from trading operations within the control after taking into account all expenses of conducting and financing them, including depreciation and interest on borrowed money, with the exclusions listed in paragraph 40, but before deducting Corporation Tax or Income Tax.

Action where Profit Margin is likely to be exceeded

63. Where:—
  (i) the reference level has been exceeded; or
  (ii) in the light of interim accounts or other evidence,
the reference level is likely to be exceeded, after taking account of seasonal and other distorting factors, abatements in allowable cost increases or price reductions should be made. The abatements or reductions should be sufficient to eliminate the actual or anticipated excess over the reference level as soon as reasonably possible, and to offset any excess

which has already arisen in a period subsequent to 30 April 1973. Abatements or reductions are however not required in respect of any excess arising in the period ending 30 April 1973.

Unit for Profit Margins

64. In calculating net profit margins under paragraph 61 enterprises may opt to use:—
either   (i) the profits of the enterprise as a whole
or   (ii) the profits of separate constitutent companies or subdivisions as defined in paragraph 26 or 27, provided that they then adhere to the basis chosen for all the purposes of the Code to which net profit margins are relevant.

65. Paragraph 61 requires that prices should be determined so as to secure that net profit margins do not exceed the reference level. A net profit margin calculated under paragraph 64 (i) will be relevant for this purpose to all the prices of the enterprise as a whole. A net profit margin calculated under paragraph 64 (ii) will be relevant to the prices of the constituent company or sub-division from the profit of which the margin has been derived.

Comment

The authors believe that the principle of relating price changes to profit rates is completely sound and possibly unavoidable. One difference between the British proposed system and PSP is that the British are not utilizing the tax system as the teeth for their enforcement. Consequently, they seem to be embarking on an endless chain of decisions which, if nothing else, should create full employment for British attorneys and accountants. With PSP, the burden for conformity to social policy would be upon the corporation, since a 100 percent tax would deny it gains for nonconformity, by transferring these gains to the society. The British firm would be responsible for making "abatements or reductions," but there seems to be no guarantee that these would necessarily reach the persons who had paid excessive prices in the first place.

READING 7

Social Responsibility of Corporations

From within the corporation as well as from outside the corporation, pressures are accumulating for an expansion of the social responsibility of corporations. Joseph W. McGuire writes:

> There are many reasons for the involvement of business in social affairs, and many reasons why this trend will continue and expand in the 1970's. Let us examine a few of these.
>
> First, the profit motive still remains extremely important. Some social ventures do, frankly, promise a return comparable to other commercial opportunities. I cannot, however, agree with those cynics who simplistically attribute all social acts by business to the profit motive: to the naive notion that economic self-interest in the traditional grand manner explains it all. . . .
>
> Second, the concept of 'profits' seems to have altered somewhat in the sixties, and will probably be broadened further in the seventies. Thus, executives have argued in recent years, that, through the efforts of their companies to improve the community they are returning to their shareholders a new type of dividend in the form of cleaner cities, safer streets, and reduced welfare and crime costs. Or that modern management is 'investing' in a society in the same manner that its traditional forerunners invested in plant and equipment. . . .
>
> Third, the vagueness of the profit concept not only weakens traditional drives and measures, but it permits other motives to assume greater importance in business. . . .[1]

Milton Friedman dissents, and perhaps his view is similar to that of many companies. He writes:

When I hear businessmen speak eloquently about the "social responsibilities of business in a free-enterprise system," I am reminded of the wonderful line about the Frenchman who discovered at the age of 70 that he had been speaking prose all his life. The businessmen believe that they are re-defending free enterprise when they declaim that business is not concerned "merely" with profit but also with promoting desirable "social" ends; that business has a "social conscience" and takes seriously its responsibilities for providing employment, eliminating discrimination, avoiding pollution and whatever else may be the catchwords of the contemporary crop of reformers. In fact they are—or would be if they or anyone else took them seriously—preaching pure and unadulterated socialism. Businessmen who talk this way are unwitting puppets of the intellectual forces that have been undermining the basis of a free society these past decades. . . .

In a free-enterprise, private-property system, a corporate executive is an employee of the owners of the business. He has direct responsibility to his employers. That responsibility is to conduct the business in accordance with their desires, which generally will be to make as much money as possible while conforming to the basic rules of the society, both those embodied in law and those embodied in ethical custom.[2]

Contrary to Friedman's viewpoint, Ralph Nader and his associates are trying to get the public represented on corporate boards. Corporate officials espouse concern for environment, discrimination, employee welfare, and other such topics. There is a major conflict here, however, and, as Friedman points out, this stems from the goal to maximize profit or at least achieve the minimum acceptable level of profit.

If corporations improve the safety of their products, for example, they are apt to reduce profits unless they raise their prices. And they can't raise prices unless all companies in that production sector also raise their prices to a similar degree. Some corporations may wish to improve their products or reduce their pollution but others may not be so socially concerned. Even where the managers have some social conscience, however, their concern for high profits may take first priority. They may feel that their profits must be at least as high as their competitors' profits. In this case, unless all act, no one is likely to act.

One way to achieve action is via laws and their enforcement, which apply to all corporations alike. Unfortunately, such laws must be very complex and detailed to cover all circumstances. An indirect approach which encourages corporations to be more responsible in conformity with each corporation's unique circumstances is preferable as it is more flexible and decentralized. One example of this approach is PSP (Profit Stabilization Principle), and we shall discuss this in more detail later. First, let us examine the problem in a more concrete way.

Once I (J. W. Leasure) met a couple at some meetings regarding urban growth and pollution. She was opposed to continued growth and concerned about the environment. He worked for some large industry in a professional capacity. After he told me the name of his company I recalled how their factory had neglected landscaping the grounds around their building. There was a high wire fence and not one bush or tree was present to soften the appearance of the wires. Furthermore, this building was located in an area that the city was landscaping for a park because of its scenic potential.

I suggested to this man that some trees could be planted, as I assumed he would be open to such an idea given his presence at environmental meetings. But immediately he pointed out that trees, bushes, and their care cost money. He was indignant that I would expect his company to reduce its profits just to beautify the area for the benefit of the passersby.

The attitude that profit is first and foremost may be softening somewhat, but if so, the process is very painful and slow. Given the conditioning of our system in favor of competition, and the accumulation of wealth and power as symbols of status, only a few people find it possible to act independently of these social and business pressures. People can be exhorted to be more responsible socially, but what are the rewards for being socially responsible in our system? All the rewards lie in the other direction. Why don't companies announce the fire hazards of their products, for example? Obviously, it would reduce their sales and profits. Why do companies wait as long as possible to correct their pollution? As long as it is financially advantageous to wait, they do so. But when public pressure mounts and lawsuits are threatened, then the financial benefits of waiting are reduced. In fact, threatened lawsuits can make it more profitable to take corrective measures.

It seems that social responsibility is acquired much more quickly when one's pocketbook is involved. With economics as the basic consideration motivating people to act, then other factors can reinforce and interact with economic factors. In fact, people sometimes even seek to explain their behavior solely in terms of moral principles, religious beliefs, or concern for others, even though economic considerations are their ultimate concern.

We assume that in our society, selfish economic considerations are dominant. The question is how to work within the spirit of selfishness to achieve more social responsibility. The answer to this question is one aspect of what we have attempted through PSP.

PSP puts an outer limit on profit. It sets a limit on our acquisitive spirit by providing no more economic rewards after a certain point. But selfish interest is allowed to prevail up to a point, and that point is reached at a rate of return after normal taxes of around 10 percent. After that, all additional profit is taxed at the rate of 100 percent. At that point, therefore, there are no more financial rewards for selfish versus social behavior. Consequently, the companies have an incentive to develop their social conscience.

Let us refer back to our earlier example regarding landscaping. When a company's profits are generating the allowable maximum, rather than pay any additional taxes they may as well raise costs and reduce profits by improving the environment. This would also apply to antipollution measures. When there is social pressure to reduce pollution, then the company benefits from doing so in terms of public relations, and it is not hurt financially if profits are pushing the maximum.

With this type of restraint on accumulation of wealth, it appears that the development of corporate social responsibility could proceed at a much faster rate than it is doing at the present time.

One other aspect of the need for more social responsibility stems from the sheer size of many corporations. Over the years, the number of corporations dominating the economy has grown smaller and smaller. Simultaneously, their power has grown larger and larger. Increased power should be accompanied by a keener sense of responsibility or by social restraints on behavior, if we are to avoid potential injury to many people. It is becoming more apparent that what is good for General Motors, for example, is not always good for the rest of us.

When a company is very small, it cannot do extensive damage to society since its sphere of influence is limited. Consumers, for example, can choose where to purchase things and have some choice about where they can find employment. But when a few companies dominate an industry, either directly through ownership or indirectly through interlocking directorates and other alliances, the picture is different. In that case, a decision made by relatively few people regarding prices and production has repercussions throughout the economy and leaves no one unaffected. For example, if the price of steel is increased, then every company using steel will be under pressure to raise its prices.

Corporations must consider the social implications of their behavior. But unless criteria and restraints are imposed on all corporations, one group cannot be expected to act in the interest of society while the other group acts in self-interest, particularly when the latter group is rewarded with higher profits. As a consequence, we have proposed PSP.

CHAPTER 4

READING 8

Guaranteed Income Plans and Some Alternatives

Many people are poor and able to work but unable to find jobs because they lack skills currently in demand. Others work but are poor because their limited skills do not command much money in the market. The great majority of the poor, however, are unable to work for reasons such as the presence of small children, age, poor health, and mental or emotional problems. All these are the people who need some kind of guaranteed income programs.

In the more distant future, as the number of people seeking work increases and the number of jobs available, particularly of the unskilled and semiskilled type, fails to keep pace, then such people will also need income supplements. Jobs could be shared, the work week could be reduced, the government could employ more people—all these measures would help distribute work. But as the growth rate of the economy slows down for reasons of environment and scarcity of resources, then some minimal income guarantees are mandatory for those who cannot find remunerative work.

In order to understand a guaranteed income plan, let us use an example based on extreme optimism. Assume a guaranteed minimum of $5,000 for a family of four with incentives thereafter up to an income of $6,500. What this means is that if the family earnings are zero, it will receive $5,000. The marginal (additional) tax rate on all private earnings up to $3,000 is 50 percent. Under most present welfare schemes, the marginal tax is 100 percent; in other words, the family's welfare payments are reduced by the amount of its earnings. The more the family earns, the more it is taxed, with total income remaining constant so long as the family is on welfare. No incentive is given for working.

In the example presented here, the total income, consisting of private earnings plus supplementary payments by the government, is limited to $6,500. When a family earns $6,500 then all income supplements cease. When private earnings rise from $3,000 to $6,500, the government takes an ever larger proportion of earnings in taxes while maintaining the $5,000 guaranteed minimum payment. Total income, as a consequence, remains at $6,500 in this range. Table 9 outlines the relevant data.

In this example, a family earning between $3,000 and $6,500 would receive the same total income of $6,500. It could be argued that there is no work incentive in this income range since the total income does not increase with rising private earnings. This is the

TABLE 9

Guaranteed Income up to $6,500

Private Earnings	Taxes Paid on These Earnings	Income Retained	Guaranteed Minimum	Total Income
0	$ 0	$ 0	$5,000	$5,000
1,000	500	500	5,000	5,500
2,000	1,000	1,000	5,000	6,000
3,000	1,500	1,500	5,000	6,500
4,000	2,500	1,500	5,000	6,500
5,000	3,500	1,500	5,000	6,500
6,000	4,500	1,500	5,000	6,500
6,500	0	6,500	0	6,500

consequence of putting an upper limit of $6,500 on total income when the government is supplementing it. A counterargument in favor of this scheme, apart from the need for some upper limit, is that in the range from $3,000 to $6,500 people will work regardless of the reduced incentives in the hope that evenutally they can raise their private earnings to more than $6,500 and at that point begin to raise their total income. There could be many variations of this scheme such that total income would rise continuously the more one worked.

A plan much discussed is the negative income tax whereby people receive money *from* the government (negative taxes) instead of paying money *to* the government when their income falls below a certain level.

One variation of this proposal is the family assistance plan proposed by Mr. Nixon several years ago. It is a type of guaranteed income with an incentive to be available for families with children. Most likely, some form of this will be passed by Congress and approved by the president in the 1970s. The bill was known as H.R.1 and guaranteed a minimum of $2,400 and paid benefits up to an income of $4,200. Its cost would be much less than in the earlier example, of course. Table 10 gives the figures covering that plan, as calculated from provisions of H.R.1.[1]

TABLE 10

Relationship under H.R.1 between Earnings, Welfare Benefits, Taxes, After-Tax Income, and Marginal Tax Rate for a Family of Four, at Selected Earning Levels in Dollars

Annual Earnings	Welfare Benefits	Income Before Taxes	Social Security Taxes[a]	Individual Income Taxes	Income After Taxes	Marginal Tax Rate
0	2,400	2,400	0	0	2,400	0
720	2,400	3,120	39	0	3,081	5.4
1,200	2,080	3,280	65	0	3,215	72.0
1,800	1,680	3,480	97	0	3,383	72.0
2,400	1,280	3,680	130	0	3,550	72.0
3,000	880	3,880	162	0	3,718	72.0
3,600	480	4,080	194	0	3,886	72.0
4,200	0	4,200	227	0	3,973	85.4
5,400	0	5,400	292	155	4,953	18.3
6,000	0	6,000	324	245	5,431	20.4

a. The employee payroll tax rate of 5.4 percent, established in H.R.1 for 1973 and 1974, is assumed in these calculations.

©1972 by the Brookings Institution, Washington, D.C.

Another approach to income maintenance with respect to children is a plan whereby children from lower-income families are paid for going to school. This would be a variation of the G.I. Bill extended to earlier ages and restricted to the poor. Children and parents would be rewarded for doing something socially approved and generally desired.

This proposal is outlined below in an article by the authors which appeared in the *Los Angeles Times,* January 11, 1968. The figures refer to 1966 dollars. To adjust for inflation since then, we would have to increase the dollar estimates by about one-third. We think the idea still merits consideration as an alternative to, or as an adjunct to, income maintenance programs for the school-age population. Just as under the negative income tax there is an incentive to help oneself by working, in this plan the incentive is provided for helping oneself by attending school.

THE POVERTY TRAP
PAY-FOR-SCHOOL MIGHT OPEN THE DOOR

Many specific economic proposals have been put forward to remove the more undesirable effects of poverty, including income maintenance or a negative income tax or "children's allowances." We propose an alternative to such benefits for that segment of the non-institutionalized population which is pre-school and school age, including college.

Our proposal is that persons should be compensated not for being poor as a negative income tax would do, nor for having children as children's allowances suggest, but for doing that which is likely to benefit themselves and society concurrently, i.e., going to school.

We are urging the Department of Health, Education and Welfare to seek legislation to establish a kind of G.I. Bill beginning at birth for children whose parents fall in the *lowest 20%* of earning groups. Such legislation would underwrite part of the board and school costs from pre-school through college for children who might otherwise follow their parents into the poverty trap.

* * *

Our approach is based first on the impressive evidence that a major source of economic growth (and the rise in the standard of living of the people of the United States) is the rising educational attainment of the population in general; and second, that the economic success of individuals is closely related to their years of quality schooling. The success of the World War II G.I. Bill for education suggests that our program, while bold, is not unprecedented.

We estimate that the underwriting of part of the room and board and all of the educational costs of the lowest 20% of the income receivers of the expected population of 1970 (using 1966 dollars) is approximately $10 billion, a figure well within the reach of the U.S. economy. This estimate includes both direct payments to parents and scholars and the public costs of expanding the educational system to accommodate the increase in years of school attendance of the lowest income group.

These estimated costs are based on the following figures. The mother, parents, or guardians of a child in the lowest fifth income group (regardless of whether on welfare or merely poor) would receive $25 a month per child from the child's birth to the age of three, contingent only upon the taking of the child for regular health checkups, well baby clinics and the like, where provided.

When the child is three, those responsible for the child will continue to receive this monthly allotment if they see to it that the child attends kindergarten until the age of six, when he enters the first grade. Evaluation of coverage would be annual, in regard to income, and periodic in terms of school attendance, as was the G.I. Bill.

Throughout elementary school, even though school is compulsory, the parents will receive $25 a month per child in school. Beginning with high school, however, each student established and continuing in the program (and not the parents) will receive $50 a month for 12 months of each year in which he does satisfactory school work.

The 14-to-18-year-olds can pay their parents a certain amount each month for room and board, or if necessary, live with relatives and pay them for room and board. This is an important aspect of the program, since it provides an attractive alternative to dropping out of school prematurely.

If such a student can do college or other approved training successfully, he or she will receive $75 a month for room and board while in college, trade school, or business college. These payments of $25, $50, and $75 are the proposed allotments in 1966 dollars. Should inflation raise the costs or should society think the allotments too high or too low, payments could be altered.

Since these payments are really incentive payments rather than a complete underwriting of costs, it is anticipated that the scheduled amounts will be adequate to persuade and enable the lower economic groups to attend school in the same pattern as middle income groups. In economists' terms, the plan would raise the opportunity cost of leaving school. If the lower fifth income group attended high school and college in the same proportion as the national white population did in 1966, the cost of the scheduled payments to parents and individuals would be $6.5 billion.

The public educational costs of these additional numbers of persons in pre-school, high school, and college would add another $3.5 billion to the cost. (School attendance at the elementary level would not be affected.) Educational costs of approximately $1 billion would be incurred for a head-start program for the ages of three to six for 1.6 million children at an annual cost of $619 a student; and $2.5 billion for high school and college costs based on a public school estimated expenditure of $893 for high school and $2,300 for college would also be needed.

The total costs for the year 1970 of $10 billion would vary in subsequent years only as educational costs rise, as population increases, or as monthly allotments are altered.

We believe that an enormous payoff to society is implied if this approach is used, since both the economic usefulness and the quality of life of a sizable segment of the American population would be directly affected. While the program is manipulative, individuals are paid only for doing that which is deemed appropriate to the economy, and it is in the spirit of our society to give economic rewards for economic contributions.

In another study, we found that group returns for raising the educational level of the low-school-attending group (which coincides with the low income group) were high enough to repay fully the additional investment.

Comparing the education costs and educationally related earnings of three different income groups through the use of actuarial tables, we could identify the benefits which would accrue to groups of individuals in the form of higher income as a result of their being able to make greater contributions to the economy because of better training. Beneficiaries of rising educational levels eventually repay a society having a progressive income tax, out of these increased earnings. Predictably, some of the beneficiaries of our proposed plan might someday complain of the burden of public expenditures on education, even as many persons educated under the World War II G.I. Bill do now.

Furthermore, we believe that society will be repaid in other ways if a school attendance support program is implemented. National income will continue to rise, and the social costs of crime, juvenile delinquency, and institutionalization which are apparently related to poverty and to poor school attendance, will decline.

Ironically, many American economists, willing to advise developing countries to emphasize education, are reluctant to propose the same breakthrough patterns in our own underdeveloped areas of human investment. While this proposed program provides no panacea, particularly in that it is directed only to the young, we consider it an essential part of any real escape from the poverty trap.

Relationship between Welfare Payments, the Cost of Welfare Programs, and the Level of Employment

Evidence is mounting that the most constructive guaranteed-income plan for those able to work is the guarantee of a job, a matter discussed in more detail in reading 9. The evidence usually takes the form of relating changing costs of existing welfare programs to changing levels of employment.

Governor Reagan of California became a national hero to certain tax groups by an attempted, though partly abortive, effort to tighten up welfare regulations by denying eligibility. After the passage of California's Welfare Reform Act of 1971, the welfare caseloads in Los Angeles County did decline dramatically and the governor claimed credit. However, Ellis P. Murphy, Los Angeles County welfare director, claimed that the biggest factor—perhaps 60 percent of the decline—was due to the drop in unemployment rates. For example, the county general-relief load has been cut almost in half, from 27,000 to 14,000 persons. The first factor in the reduction of welfare caseloads is the rising employment level; the second, the shrinking of the average welfare family from 3.2 persons to 2.3—"obviously due to abortions and family planning."[2]

READING 9

Issues in the Solution of the Problem of Unemployment:
The Federal Government's Role as Employer of Last Resort

The economist has traditionally distinguished between frictional unemployment, associated with temporary changes in jobs to reflect normal adjustments of both market and individual goals, and cyclical unemployment, which is associated with "business recessions" or "failure to achieve the normal rate of growth." Increasingly, the government is blamed for cyclical unemployment, since most economists no longer expect prosperity without government manipulation of fiscal and monetary controls. Frictional unemployment is the lifeblood of necessary adjustments in a dynamic economy, but cyclical unemployment represents a "failure." The United States has failed so consistently in achieving desired employment levels that, since the fifties, American economists have found it convenient to talk about "structural" unemployment, variously attributing such unemployment to characteristics (youth, lack of skills) of the labor force itself or to the labor market (technological displacement).

While "full employment" is in one sense a spurious goal which evokes to some the image of a slave state, currently it is the "liberal" goal in America and the conservative goal in other market-oriented economies. The reason is, of course, that one's standard of living ordinarily depends on working and that even in the goods sense, nothing produced is nothing consumed for society as a whole. If levels of living generally, as well as individually, are to rise, then work-effort is essential.

Looking around the world, one sees that the problem of unemployment has many aspects:

1. Some countries apparently have too few resident workers —for example, postwar Germany, which has depended on workers from Italy and more recently from Turkey and Yugoslavia.
2. Some countries apparently have too many workers—for example, most of the less developed countries, which exhibit not only classic unemployment but also underemployment, with many workers doing what a few could do, particularly in the agrarian sector.

The United States, regionally viewed, has both these characteristics. In the Southwest many employers are dependent on alien workers. In the Southeast, particularly Appalachia, underemployment is endemic.

Because of the structural and regional nature of unemployment in the United States, the Congress has attempted programs which would treat these sources of unemployment directly. The Emergency Employment Act of 1971 proposed the principle of the government's acting as employer in response to unemployment, through federal provision of grants to local governments. One of the questions which this experimental legislation posed was whether there would be a multiplier effect of such government-created employment. The following case study reviews the first year's experience in one community with this question in mind.[1]

EVALUATION: AN EXPERIMENT IN HIGH IMPACT FUNDS
ON TOTAL UNEMPLOYMENT: INITIAL Year

The San Diego Public Employment Program (PEP) represented an effort to determine what would happen if funds sufficient for hiring 10 percent of the unemployed came into the community. The total of Emergency Employment Act funds promised to San Diego exceed $20 million, and the number of jobs created reached a peak in May with 1,227 city-administered slots, 1,307 county-administered and more than 266 state-administered, for a maximum of 2,800. As of September 1, 1972, more than $13 million in participants' wages had been expended.

Any investigation of the impact of the addition of so many jobs and so much money must include an understanding of the nature of the local economy, the changes which occurred in it during the year, and other related factors.

Role of Government in San Diego County

As recently as 1958, 19.2 percent of San Diego County's civilian employed workers were engaged in federally financed defense-oriented activities. In manufacturing in that year, eight out of ten workers (77.7 percent) were in defense-related employment. After this peak, adjustments were made until in 1968 only 10.2 percent of civilian employment and only 56 percent of manufacturing employment was in defense-oriented jobs.[2]

In aircraft and ordnance alone, employment levels dropped from 32,800 jobs in 1968 to 21,200 in July 1972.[3] Thus San Diego has made substantial adjustments and was continuing these adjustments even before the onset of the recession in early 1970.

Agriculture

San Diego employment is mainly in nonfarm undertakings. Yet agricultural employment, which represents only 21 percent of the total employment, nevertheless accounts for sizable fluctuations in the total civilian employment levels. For example, the number of persons employed in agriculture from August 1971 to August 1972 ranged from a low of 9,200 in January to a high of 12,100 in May. Unfortunately PEP did little for this source of unemployment, since the cumulative total of "Seasonal Farm Wage Workers" reported as participants was only 6. Also, because San Diego is so near the Mexican border, many American seasonal farm workers are of limited English-speaking ability. The cumulative total of such participants (35) was minimal.

Manufacturing and Technological Displacement

Although manufacturing provides only 14 percent of wage and salary civilian employment in San Diego, it is also the source of an overall decline of 4,200 jobs. At least some of this is due to federal retrenchment. Two industries are accepted as primarily sensitive: aircraft and ordnance on the one hand and shipbuilding on the other. The changes in employment levels during the contract year of these suggests this possibility: the probable maximum of 2,800 PEP jobs could not have offset the decline of 5,500 jobs lost in aircraft and ordnance and shipbuilding. (See table 11 for a summary by months of these and other industries.)

Since technological displacement was an important source of unemployment during the PEP year, the question of whether PEP had any impact on this type of unemployment is important. The record here was better than in the case of agriculture. City had 110 cumulative participants who had been technologically displaced as of September 1. County had 196. This represents more than 300 technologically displaced workers who were served by PEP.

On the other hand, in actual numbers, the 5,200 job increase in government employment, most of which was PEP, more than offset the overall decline in manufacturing of 4,200, and this was true in July, after the peak of current PEP employment had been passed. Clearly, part of the local adjustment was from federally financed manufacturing to federally financed PEP.

It is thus reasonable to claim that PEP prevented unemployment from rising to levels even higher than those that were experienced. For example, during the months of peak PEP employment, San Diego nevertheless suffered its highest seasonally adjusted unemployment rates of the year. Take June, when aircraft ordnance had moved beyond the effects of a strike, and shipbuilding had even picked up a bit from an April low, when government employment was at maximum both at the federal and local levels. The economy was expanding moderately, yet the seasonally adjusted unemployment rate remained at 6.3 percent, partly because of the influx of young laborers into the market. This being a normal condition of the June market, one can calculate that with 2,800 fewer jobs (i.e., without PEP) unemployment would have been at 37,000, rather than 34,200, and the unadjusted rate at 7.4 percent at the least, rather than 6.8 percent.

The contrary claim, that PEP in a high impact area should have

TABLE 11
San Diego Labor Market Summary During PEP Initial Year, 1971-72

Net Change Aug. 1971 to July 1972		Aug.	Sept.	1971 Oct.	Nov.	Dec.	Jan.	Feb.	1972 Mar.	Apr.	May	June	July
					(in thousands)								
Lab FC	+13.2	484.4	484.4	483.8	484.1	484.1	482.6	488.7	490.4	490.0	490.6	499.8	497.6
Unempl	+ 1.9	29.0	26.4	26.3	26.5	27.3	29.9	32.8	32.2	30.1	30.9	34.2	30.9
Rate (seas adj)		5.9%	5.8%	6.0%	5.5%	5.9%	6.0%	6.2%	6.3%	6.7%	6.6%	6.3%	5.9%
Rate		6.0%	5.5%	5.4%	5.5%	5.6%	6.2%	6.7%	6.6%	6.1%	6.3%	6.8%	6.2%
Tot Civ Em	+11.3	455.4	458.0	457.5	457.6	456.8	452.7	455.9	458.2	459.9	459.7	465.6	466.7
Construction	+ 2.5	21.7	23.3	22.9	23.1	22.6	22.3	22.7	23.2	23.5	21.3	23.8	24.2
Trade	+ 3.6	90.6	90.9	91.2	91.8	94.7	92.1	91.9	92.0	92.3	92.9	93.5	94.2
Serv	+ 3.8	79.6	79.8	77.2	76.7	76.8	76.9	77.3	77.7	77.9	78.3	80.5	83.4
Govt	+ 5.2	100.7	102.4	104.2	105.1	105.8	106.7[1]	107.5	107.8	108.2	108.4	109.3[3]	105.9
Fed	+ .7	33.7	33.4	33.5	33.6	33.6	33.5	33.5	33.5	33.6	33.7	33.9	34.4
S&L	+ 4.5	67.0	69.0	70.7	71.5	72.2	73.2	74.0	74.3	74.6	74.7	75.4	71.5[4]
Mfg	- 4.2	61.7	61.5	61.2	60.6	56.8	56.6	58.4	57.7	57.1	57.1	57.3	57.5
Aircr Ord	+ 4.2	25.4	25.4	25.3	24.9	21.6	21.8	23.7[2]	23.2	22.4	22.0	21.8	21.2
Shipbldg	- 1.3	4.9	5.0	4.4	4.1	4.0	3.6	3.5	3.3	3.2	3.4	3.5	3.6

		Aug.	Sept.	Oct.	Nov.	Dec.	Jan.	Feb.	Mar.	Apr.	May	June	July
PEP Level[5]	2404	0	65	118	613	1550	Level	2305	2410	2503	2564	2506	2516
City	1169	—	na	na	282	734	City	1153	1172	1219	1257	1227	1255
County	1235	—	na	na	331	816	County	1152	1236	1284	1307	1279	1261

1. "Most of the jobs added to the public payroll over the past 3 months have been federally funded Public Employment Program positions within local government," *San Diego Labor Market Bulletin*, January 1972.

2. "Labor-Management Dispute was settled and employment rose by 1.9 to 23.7, well below the prestrike total of 24.9," *San Diego Labor Market Bulletin*, February 1972.

3. Maximum of PEP Currents, i.e., before freeze and before summer schools layoffs.

4. "Many classified schools personnel were laid off for summer vacations." *San Diego Labor Market Bulletin*, July 1972.

5. Net change to 9/1/72.

a multiplier effect and thus add more jobs to the local employment rolls than its own slots alone, is more difficult to substantiate. Actually, there was a steady advance in total civilian employment over the PEP year.

Most advances were in the trade and services sector, which together added about 7,400 jobs to the total civilian employment. In San Diego, these industries are sensitive to both local personal income levels and also external income levels, because of the importance of the tourist trade. As indicated below, PEP was an important personal income stabilizer locally, and probably also in Los Angeles County, which is the source of many San Diego tourist dollars.

How many of the 11,300 additional jobs, other than its own 2,500, can be attributed to PEP is indeterminate. However, it was well recognized that PEP contributed to optimism about the local economy in these positive ways:

1. Personal income was maintained at an initial level roughly $10 million higher, with prospects that this would continue for another year. (This is .5 percent of City's disposable income.)
2. City and County had a substantial incentive in continuing to maintain their own spending levels because of contract obligations.
3. The marginal propensity to consume of a PEP labor force which was formerly totally unemployed, must approach 100 percent of take-home pay. Most PEP workers receive wages of less than $4.00 an hour, and one-third were formerly disadvantaged. The effects of such a group spending-effort will depend on the group's size and its past debts. But if one considers only the reported expenditures for participants wages for the city and county PEP year, then the total wage of over $12 million, minus taxes, debts, and insurance, must yield an injection of at least $9 million. Since the multiplier depends not only on the marginal propensity to consume but also on the marginal propensity to import from other regions (which must be rather high for San Diego), then probably the

multiplier is less than 2. Let us assume, however, that the income multiplier is 2; in that case, one injection of PEP's $9 million would become an $18 million addition to personal income, over the period of one year.

If we *could* assume that each additional $4,000 creates another job, then this rate of spending would account for 2,250 of the 7,400 new jobs in trade and services. But more likely this additional $18 million would do nothing of the kind. The economic report of the governor of the state of California for 1972 projected that even an increase of personal income of $4.6 billion would not be associated with any additional jobs.[4] This points up the necessity of distinguishing between an income multiplier and a "jobs" multiplier. As we know of no regional work being done on this, we can only guess that the divergence of income and employment changes is partly due to technological progress, which accounts for rising incomes *without* increases in the demand for laborers.

On the brighter side, Congress should note that in San Diego their brainchild EEA succeeded in creating 2,500 jobs with the small expenditure of $12 million.

Conclusion: High Impact Funds

While a specific multiplier or numeral is not assignable to the PEP High Impact, there is reason to believe that the income stream was the beneficiary of an additional $18 million (participants' wages plus secondary expenditures) because of the PEP participants' high marginal propensity to consume.

The marginal propensity to consume of the whole community and its willingness to invest were also affected positively by this experience. Some additional jobs in trades and services may have resulted. Maintaining personal income was doubly important in view of substantial layoffs in defense-industry employment and widespread exhaustion of unemployment benefits.

Furthermore, there is evidence that without PEP, the unemployment rate would have been at least one percentage point higher.

Presumably the impact would have been on those persons who were PEP participants.

Any other federal expenditure of a similar sum would probably have created many fewer jobs and possibly have had some negative effects on community confidence. Hence, the opportunity cost of using this $12 million in some alternative way is very high. To put it differently, in order to dampen the unemployment rate in San Diego County as much as PEP did, alternative federal expenditure patterns would have required larger amounts of money, possibly two or three times as much.

CHAPTER 5

READING 10

Optimal Population (A Spaceship Approach)[1]

Many variables are suggested as inputs for determining the optimal population, i.e., that population size which maximizes the value of some measure of well-being. "Optimal" can be regarded in an economic sense, where one maximizes income per person. Or it can be regarded as a welfare concept, where one maximizes welfare. In this case welfare can be made dependent not only upon income per person or the utility derived from income, but also upon the utility or satisfaction that is derived from children. Some people may wish to determine that population size which maximizes some other variable such as health, clean environment, or religiosity. We would like to add a constraint on any optimizing function with respect to population. This constraint is the preservation, intact, of Spaceship Earth. Kenneth Boulding initiated this concept and we would like to apply it to the concept of optimal population.[2]

Regardless of what is being optimized—income per person, welfare, utility—there should be a constraint imposed which preserves the wealth of the earth. That is, we must replace not only the physical stock of capital or manmade means of production that are used up in the course of our productive flow of goods and service. We must also restore natural capital, so that the stock of wealth, including the physical resources, will remain undepleted. This means that air and water must be purified, and objects which use nonrenewable resources, such as stocks of metals, must be recycled or stored for future recycling. Renewable natural resources, such as forests and stocks of fish, must not be allowed to disappear.

This approach also implies that the destruction of a natural resource now, which may not mean a shortage of that material until some date in the distant future, is regarded nevertheless as tantamount to a scarcity or cost at the present time. When a future cost or deterioration is regarded as the same as one that occurs here and now, economists say that there is a zero discount rate. We suggest the use of a zero discount rate with respect to the earth's natural wealth. That is, damage which does not manifest itself significantly until fifty years from now is, nevertheless, treated incrementally. Corrective measures should be taken each year to offset the gradual deterioration. This is how we deal with physical capital.

If society wishes to achieve that population size which maximizes income per person, for example, then the income per person must be interpreted to be net income after restoration of physical capital *and* natural capital, so that the physical and natural wealth are unchanged. Today we consider net national product only, which means gross national product minus depreciation, which is the wearing out of buildings and machinery. The new concept would require a deduction for the wearing out of the environment as well as the physical capital.[3]

Spaceship Earth should remain undiminished and unscarred. The value of whatever one is maximizing should be a net figure after the restoration of nature so that the physical world remains as beautiful and rich as ever.

Notes

Chapter 1

1. John Bates Clark, *Distribution of Wealth* (New York: Kelly & Millman, 1956). This book was first published in the 1890s.
2. Cf. David Laidler, "The Phillips Curve, Expectations and Incomes Policy," in *The Current Inflation*, ed. H. G. Johnson and A. R. Nobay (New York: Macmillan Co., 1971).
3. Neil Chamberlain and Donald E. Cullen, *Labor Sector*, 2d ed. (New York: McGraw-Hill, 1971), p. 8.
4. Herbert Stein, chairman of the President's Council of Economic Advisers, said in April 1973, "We have not come to the end of Phase 3, so I believe that a judgment at this time would be premature. We will not be able to evaluate Phase 3 until many years after it is dead and we can see what its legacy has been. And, of course, what I have said about Phase 3 is true also about the whole (wage-price) controls program and about its successive phases." *(Los Angeles Times*, April 29, 1973.)
5. Cf. Walter Adams and J. B. Dirlam, "Big Steel, Invention, and Innovation," *Quarterly Journal of Economics* 80, no. 2 (May 1966): 167–89.

Chapter 2

1. Robert Barckley of the Department of Economics, California State University, San Diego, has suggested that a preferable index of concentration might be one involving sales *and* value-added. Differences in vertical integration and methods of distribution give different concentration ratios when these are based on sales versus value-added. Also, concentration ratios should be published for all large firms—or firms with sales over ten million dollars a year—and not just the top four firms as is now done by the census bureau. Our plan would also require public knowledge as to which firms were governed by the price and profit constraints.
2. Again, Robert Barckley has suggested that the price increases should be permitted on a quarterly basis with final end-of-year adjustment. An alternative would be to tie the degree of price increase to the change in profit rates in some way, so as to avoid a large price increase in one jump. Smaller and more frequent

changes are more desirable, in his opinion, and might help prevent any speculative behavior.

One way to incorporate this excellent suggestion would be to let prices rise slightly if, for example, the profit rate fell from 10 to 8 percent. But if the rate dropped from 8 to 5 percent, a much larger price increase might be permitted.

3. We are grateful to Charles F. Dicken and George Babilot for suggesting this modification.

4. Cf. Lester C. Thurow, "Income Distribution Changes in the 1960's," in *Poverty in Affluence*, ed. R. E. Will and H. G. Vatter, 2d ed. (New York: Harcourt, Brace & World, 1970), p. 55.

5. Joseph A. Pechman has an excellent review of the issue of the shifting and incidence of the corporation tax in *Federal Tax Policy*, rev. ed. (Washington, D.C.: Brookings Institution, 1971), pp. 111–14.

Chapter 3

1. This is particularly evident since the president announced a revision of the depreciation rules which he believed would "provide greater incentive for business to invest in capital equipment." *Economic Report of the President*, February 1971, p. 91.

2. Sweden has achieved this stability of growth at about the 4 percent level and of unemployment levels under 3 percent for some years but not, unfortunately, without inflation. Cf. Robert J. Flanagan, "The U.S. Phillips Curve and International Unemployment Rate Differentials," *American Economic Review*, 63, no. 1 (March 1973): 114–31.

3. Japan's cultural pattern of one lifetime employer for a worker has inhibited neither economic growth rates nor rising labor productivity. Growth rates have typically passed 9 percent in recent years. With 1965 as a base year (=100), labor productivity was approaching 190 by 1970. Cf. *Japanese Economic Yearbook*, 1970.

4. John Maynard Keynes, "Economic Possibilities for Our Grandchildren," in *Essays in Persuasion*, (New York: Harcourt Brace, 1932), pp. 369–70. It is interesting that this essay was first published in 1930, some years before the first publication of Keynes's *General Theory*. This quotation is becoming almost as famous as Keynes's "In the long run we are dead." Cf. Joan Robinson, *Freedom and Necessity* (New York: Random House, Vintage Books, 1971), p. 117, and Warren A. Johnson and John A. Hardesty, *Economic Growth vs. the Environment* (Belmont, Calif.: Wadsworth Publishing Co., 1971), p. 192.

5. Cf. Raymond Floren, Jr., J. William Leasure, and Marjorie S. Turner, *Public Benefits Associated with Public Expenditures in Education: A Demographic Approach*, (San Diego, Calif.: Institute of Labor Economics, San Diego State College, 1969).

6. Evalyn Segal, a research behaviorist in the Department of Psychology at

California State University, San Diego, has given us the following comment: "The *quantity* of reinforcement is a much less powerful determinant of behavior (rate of response) than the manner in which reinforcements are scheduled. It would seem that *some* profits (reinforcements) are necessary to maintain behavior, but that quantity and quality of behavior do *not* increase proportionately to increases in the amount of profit."

Chapter 4

1. Pierre Teilhard de Chardin, *The Phenomenon of Man* (New York: Harper & Row, Publishers, Harper Torchbooks, 1961), p. 258.
2. Adam Smith, *The Wealth of Nations* (New York: Modern Library, 1937), p. 4.
3. See Thomas D. Crocker and A. J. Rogers III, *Environmental Economics* (Hinsdale, Ill.: Dryden Press, 1971).
4. Irving Bernstein, *Lean Years* (Cambridge, Mass.: Riverside Press, 1960) p. 241.
5. Suppose that the $7 billion proposed tax cut results in only $2 billion of investment over and above what would have occurred anyway. This is a generous estimate given the condition of the economy in 1972. Therefore, additional investment spending is $2 billion. Now if the marginal propensity to spend, referred to as MPS, (for both consumption and investment) for the economy as a whole is 2/3 after allowing for taxes, corporate savings, etc., the multiplier will be $\frac{1}{1-\text{MPS}}$ or $\frac{1}{1-2/3}$ which is three. The ultimate increase in spending or income, therefore, is $6 billion ($2 billion times 3), given this kind of tax cut and our assumption about the initial increase in investment spending. Most of this type of tax cut, at the present time, is simply a transfer payment from the poor to the rich.

Now let us suppose that the government proposed a $7 billion tax cut, the same magnitude as the previous example, but this tax cut is to apply to the very poorest group in the population. For example, all income taxes could be eliminated up to some minimum income level. These are the people who will spend the most out of their additional take-home pay. In fact, they will probably spend all of it for more consumption because they are presently hungry, poorly clothed, and even more poorly housed. So, given our assumption that they spend all of it, the initial increase in spending is $7 billion, the amount of the tax cut. This figure is then to be multiplied by the multiplier.

They will buy things from a cross section of the economy, however, so the estimated marginal propensity for the entire country applies at this point. Since that MPS is estimated at 2/3, the multiplier is still 3. The ultimate increase in income, however, is now $21 billion (7 times 3) and not $6 billion as in our previous example.

Since economists cannot honestly argue that the propensity to consume is unimportant, they argue the next best thing—that it is relatively stable. Consumption follows closely the income level, or more particularly, the permanent income level. This focuses attention properly, they think, on investment, which is more volatile. Some even argue that the marginal propensity to consume out of additional income is about the same regardless of the level of one's income.

To argue this is simply to accept the current distribution of income. In reality, more equal distribution of income would make for a higher marginal propensity to consume (MPC) and thus a higher multiplier. (The MPS is determined by MPC and marginal propensity to invest, but the MPC is the primary determinant of the multiplier.) It is obvious that the MPC of the Kennedys or the Rockefellers or even a $36,000-a-year professional is less than that of a mother with six children who earns $75 a week as a cleaning lady. All these arguments seem to be a little game economists play with themselves to keep the emphasis in the "right" place—on investment and production and "growth."

Under this analysis, how can anyone argue that the marginal propensity to consume is unimportant? There is widespread acceptance of the idea that the overall propensity to consume is determined mainly by the distribution of income. Greater equality of income, therefore, means higher propensity to consume. But all this gets lost in the so-called investment multiplier.

6. Studies of the San Diego Metropolitan Area have convinced George Babilot of the Department of Economics, California State University, San Diego, that "because of inequality in income distribution, not only are the poor denied private goods and an 'equal' voice in the allocation of resources in the private sector, but they are also denied an 'equal' voice in the allocation of resources for public goods as well, despite the assumption of equality in voting power. This occurs because low income recipients are less concerned with resource-using expenditures by the government to *complement* their incomes than with transfer types of expenditures to *supplement* their incomes. Higher income recipients, therefore, have greater influence in the use of resources in the public sector, since they opt for resource-using expenditures to complement their incomes."

7. Economists typically have used a concept called the Paretian Test for improvement, which would balance those who were better off against those who were worse off in some algebraic calculation. Of course this is valid if, and only if, relative positions are ignored. Given our present distribution of wealth, income, and goods, there is no justification for ignoring these relative positions in trying to make social policy.

8. Robert E. Barckley suggests that the way to make the meaning of public interest more concrete is to broaden the idea of property rights. For example, people would have a property right in the air, water, visual terrain, public services, and private consumption expenditures. "Recognition of such rights with class action suits could more effectively bring the public intent into sharp focus. This approach escapes the dilemma of the governmental bodies which represent interest groups rather than the public interest."

Chapter 5

1. If any policy is to be proposed regarding the military, it is that there be a freeze in dollar expenditures on the military at the present level. As the economy grows, the military budget remains fixed in dollars with the result that it becomes a decreasing proportion of the total budget.

2. Economies of scale refer to techniques of production where all inputs can be increased proportionately and output either increases less than proportionately (diseconomies of scale), proportionately (constant returns to scale), or more than proportionately (economies of scale).

If each input of land, labor, and capital is increased by 10 percent, and output increases by more than 10 percent, then we have economies of scale. Diseconomies of scale would mean that output increased by less than 10 percent.

Suppose some business uses ten men for each of three eight-hour shifts per day. Each man operates one machine, so there are ten machines. Each machine runs twenty-four hours a day. All the work is done in one building. Now suppose that the owner doubles the number of machines, the size of the building, and the number of men—all as a consequence of greater demand for the final product.

With twice as many men and machines it has now become possible for each man to specialize in only one phase of the manufacturing. Formerly, each person had to perform two steps in the production process. When each laborer finished one operation he had to spend some time closing down that phase and preparing for the next one. With more men, it is now possible for each man to work on only one phase, which eliminates the time formerly spent on switching from one task to another.

As a result of all this, output more than doubles. A doubling of the men, machines, and building would lead one to expect a doubling of output. As each man specializes further, however, output more than doubles; this is an example of economies of scale.

When a market for a product grows as a consequence of a growing and affluent population, production increases and economies of scale are probable. Economies of scale are the result of: a reduction of unavoidable "excess capacity," which results, for example, when ten trains are run per day instead of two on the same tracks; quantity discounts when goods are purchased on a large scale; specialization by men and machines; and the statistical laws of large numbers which make it less costly to meet all possible needs. All of these can result from an expanding market; consequently, size of population is important. (Cf. George J. Stigler, *Theory of Price*, 3d ed. [New York: Macmillan Co., 1966] pp. 153–54.)

Output has a broader context, however, as the word is used here. It must be viewed from the standpoint of net social product, which allows for social costs such as smog or pollution. These externalities are costs to society, if not internal costs to the firm. As they are internalized by taxes and laws, these costs become a part of the private costs of operation. In either case, they must be included in an overall analysis from the standpoint of society as a whole. To a large extent, smog and pollution are the result of the fixed amount of available natural resources. This brings us to the other determinant of optimal population—increasing social marginal cost.

"Social" refers to all costs of production, public and private, and "marginal" refers to the additional or extra cost involved in the production of one more unit. The obverse of increasing social marginal cost is diminishing social marginal product.

Increasing social marginal cost applies in a situation where at least one input is fixed. As other inputs, such as labor, are added, social marginal product begins to diminish at some point. In other words, extra output begins to fall. Consequently, the extra cost of an additional unit of output rises.

3. Figure 3 illustrates this relationship:

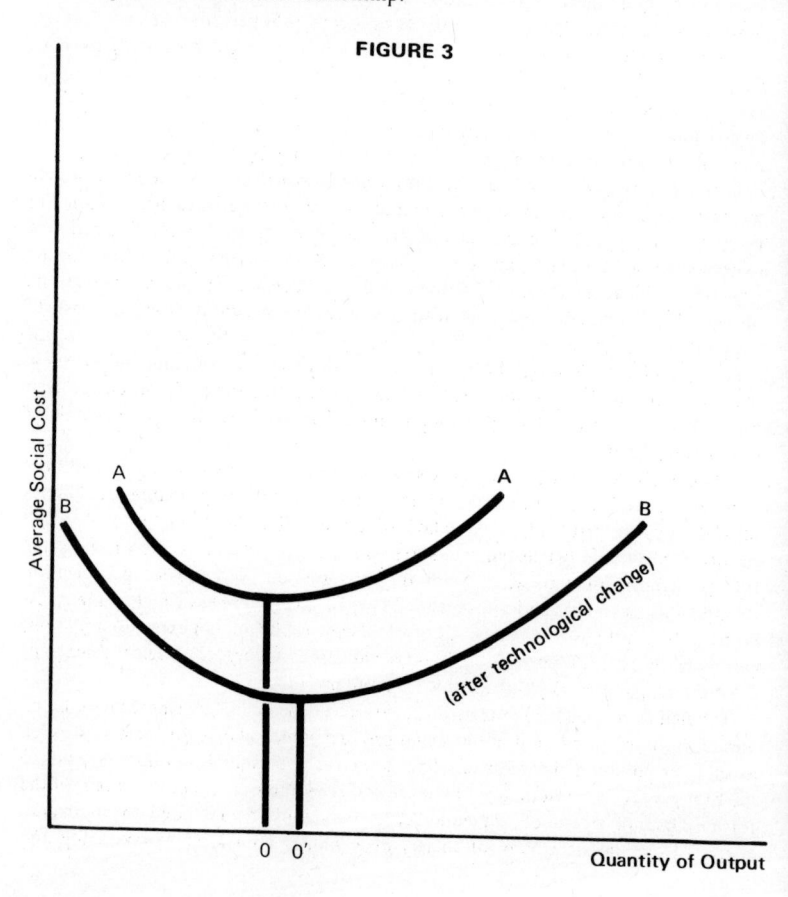

FIGURE 3

Curve AA is the average social cost curve representing the original relationship between economies of scale and social costs. At point 0, the level of production associated with a given combination of population size, age-sex distribution, and per capita income is optimal because social average costs are minimal.

This diagram allows for population and capital to increase, with a resulting rise in production. As we move along the horizontal axis, production increases. This can be the result of either an increase in population or a rise in income per capita or a combination of both. An important variable is held constant, however, and that is the state of technology. Over time, this changes, and the curve will shift. Most likely it will shift down and to the right to form the curve BB as technological change permits the more efficient disposal of wastes. If so, the "optimal" *production* level changes from 0 to 0'.

4. Even with the application of all available technology, including expected developments in the future, not all pollution is eliminated. At some point we reach a dangerous level of contamination, whether it be from noise, carbon monoxide, impure water, or simply crowding. Let point M on the average social cost curve in figure 4 be the point at which the level of pollution becomes intolerable. This point could be the result of a political decision as well as a medical one. Even with technological change, the dangerous point exists, but at a higher level of output, as shown in figure 3. Society's goal, given the desire for self-preservation, is to remain to the left of M, the dangerous level.

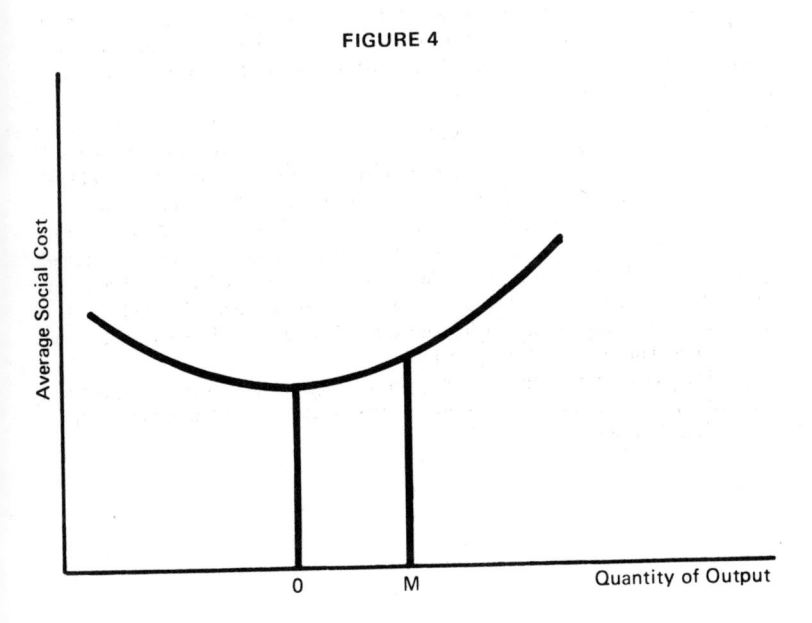

FIGURE 4

The optimal point 0 represents an infinite number of combinations of population and output per person. Therefore, there is not *one* population size associated with point 0, the optimal level of producton, but many.

We are undoubtedly to the left of M in most countries of the world today. Whether or not we are near 0, no one knows, because empirically it is impossible to measure. Nevertheless, we can draw some significant conclusions from the argument presented thus far. Given that we are to the left of M, as production increases we move closer to that point. Production, as was noted earlier, is the result of total population multiplied by output per person. Given point M, the ultimate constraint on total production, the income *per person* could rise more if population size were constant, before reaching the intolerable level of total production presented by M. For example, if 2,000 units is the maximum level of output subject to the pollution constraint, then a population of 100 versus a population of 200 yields an output per person that is twice as large. We can suppose that the population of 100 represents the size of a stationary population and that a growing population results in a population of 200.

We assume that the majority of people prefer a continually rising standard of living; this can be achieved for a longer period of time with a population that is stationary. The conclusion here is that ultimately a higher standard of living per person is possible with a stationary population.

In the analysis so far, we have not considered the effects of distribution of population on pollution and social costs. A redistribution of population could reduce average social costs of production. Development of new cities or expansion of small ones, as opposed to further growth in the large ones, might well reduce many social costs of production. Whether or not this is done will depend on the formulation of public policies in this sphere.

Let us assume that a conscious effort is made to redistribute the future population so as to reduce social costs. What will happen is that the social cost curve will shift down in a manner similar to what happened after the postulated technological change illustrated in figure 3. Repeated redistributions of the population would cause a repetition of the downward shifts. As population and production increase, however, further redistribution is without the desired results—all areas are equally contaminated. At this point, the analysis becomes the same as that described earlier. A dangerous level of pollution of one type or another ultimately will be approached. Consequently, a higher standard of living is possible with a stationary population than with one that is growing.

5. The negotiations between the oil-producing companies and oil-producing nations indicate that profits are already being subjected, at least in this one industry, to international rules.

Appendix

READING 2

1. Charles H. Hession and Hyman Sardy, *Ascent to Affluence* (Boston: Allyn & Bacon, 1969), pp. 798–99.
2. Ibid., p. 798.
3. Ibid., p. 801; original source: U.S., Congress, Senate, *Hearings on Economic Concentration*, 88th Cong., 2d Session, Part II, *Mergers and Other Factors Affecting Industry Concentration*, p. 516.
4. Philip A. Klein, *The Management of Market-Oriented Economies* (Belmont, Calif.: Wadsworth Publishing Co., 1973), p. 90.
5. Ibid., p. 93.
6. Ibid., pp. 90–91.
7. Milton Friedman, *Capitalism and Freedom* (Chicago: University of Chicago Press, 1962), pp. 121–22.
8. Ibid., p. 129.
9. Ibid., pp. 132–33.
10. Galbraith's original testimony was before the committee of the 90th Congress, June 29, 1967. Excerpts are widely available, for example in such reading texts as Paul Samuelson, *Readings in Economics*, 6th ed. (New York: McGraw-Hill, 1970), p. 222, from which these quotations are drawn.
11. Ibid., p. 227.
12. Ibid., p. 230.

READING 3

1. Friedrich A. Hayek, *The Road to Serfdom* (Chicago: University of Chicago Press, 1944), p. 83.
2. Ibid., p. 85.
3. J. M. Clark, *Alternative to Serfdom* (New York: Alfred A. Knopf, 1950), p. 20.
4. John K. Galbraith, *American Capitalism* (Boston: Houghton Mifflin, 1956).
5. Milton Friedman, *Capitalism and Freedom* (Chicago: University of Chicago Press, 1962), p. 197.
6. Ibid., p. 201.
7. John K. Galbraith, *The New Industrial State* (New York: New American Library, Signet Books, 1967), p. 316.
8. Joan Robinson, *Freedom and Necessity* (New York: Random House, Vintage Books, 1971).
9. Philip A. Klein, *The Management of Market-Oriented Economies* (Belmont, Calif.: Wadsworth Publishing Co., 1973).

READING 5

1. To be consistent with our earlier formulation of the five-year average profit rate, we should use dollar figures for profit and assets and divide the two aggregates to obtain the average rate. The result is similar to an average of the rates only if the magnitude of profit and assets does not fluctuate sharply.

READING 7

1. Joseph W. McGuire, "The Future Social Role of the Business Organization," *Review of Social Economy* 28, No. 1 (March 1970), as condensed in Kenneth G. Elzinga, ed., *Economics: A Reader* (New York: Harper & Row, Publishers, 1972).
2. Milton Friedman, "The Social Responsibility of Business Is to Increase Its Profits," *New York Times Magazine*, September 13, 1970, reprinted in *Economics: A Reader*, ed. by Kenneth G. Elzinga.

READING 8

1. Charles L. Schultze et al., *Setting National Priorities: The 1973 Budget* (Washington, D.C.: Brookings Institution, 1972), p. 192.
2. *Los Angeles Times*, Sunday, June 10, 1973, Part II. Murphy's conclusions were based on a study prepared by the county's Department of Public Social Services which held that "no change in law or policy since 1968 has had a significant systematic or sustained impact" on the welfare rolls.

READING 9

1. *The First Year Experience with Public Emergency Employment: San Diego City and County*, Institute of Labor Economics, California State University San Diego, October 1972.
2. San Diego (County) Planning Department and Chamber of Commerce.
3. *Labor Market Reports*, California Human Resources Department, Monthly.
4. Economic Report of the Governor, California, 1972, projections:

	1970	1971
Personal Income (in millions) (selected indicators):		
Wages and Salaries	60.0	62.6
Personal Income	88.8	93.4
Employment and labor force (in thousands):		
Total Employment	8036	8004
Labor Force	8555	8603
Unemployment	519	599
Rate	6.1	7.0

Appendix

READING 10

1. This paper was written by V. G. Flagg and J. W. Leasure. Professor Flagg is Research Associate in the Center for Public Economics, California State University, San Diego.

2. See Kenneth E. Boulding, "The Economics of the Coming Spaceship Earth," in *Environmental Quality in a Growing Economy*, ed. Henry Jarrett (Baltimore: Johns Hopkins University Press, 1966). See also Kenneth E. Boulding, "Fun and Games with the Gross National Product," in *The Environmental Crisis: Man's Struggle to Live with Himself*, ed. Harold W. Helfrich, Jr. (New Haven: Yale University Press, 1970).

3. This concept is further developed by William Nordhaus and James Tolien. Their Net Economic Welfare, as distinguished from Gross National Product, allows an adjustment of GNP to correct for the disamenities of modern urbanization, such as pollution. See W. Nordhaus and J. Tolien, "Is Growth Obsolete?" in National Bureau of Economic Research, *Fiftieth Anniversary Colloquium V* (New York: Columbia University Press, 1972).

Bibliography

Bernstein, Irving. *Lean Years.* Cambridge, Mass.: Riverside Press, 1960.

Chamberlain, Neil, and Cullen, Donald E. *Labor Sector.* 2d ed. New York: McGraw-Hill, 1971.

Clark, John Bates. *Distribution of Wealth.* New York: Kelly & Millman, 1956.

Clark, J. M. *Alternative to Serfdom.* New York: Alfred A. Knopf, 1950.

Crocker, Thomas D., and Rogers, A. J., III. *Environmental Economics.* Hinsdale, Ill.: Dryden Press, 1971.

Elzinga, Kenneth G., ed. *Economics: A Reader.* New York: Harper & Row, Publishers, 1972.

Floren, Raymond, Jr.; Leasure, J. William; and Turner, Marjorie S. *Public Benefits Associated with Public Expenditures in Education, A Demographic Approach.* San Diego, Calif.: Institute of Labor Economics, San Diego State College, 1969.

Friedman, Milton. *Capitalism and Freedom.* Chicago: University of Chicago Press, 1962.

Galbraith, John K. *American Capitalism.* Boston: Houghton Mifflin, 1956.

———. *The New Industrial State.* New York: New American Library, Signet Books, 1967.

Hayek, Friedrich A. *The Road to Serfdom.* Chicago: University of Chicago Press, 1944.

Helfrich, Harold W., Jr., ed. *The Environmental Crisis: Man's Struggle to Live with Himself.* New Haven: Yale University Press, 1970.

Hession, Charles H. and Sardy, Hyman. *Ascent to Affluence.* Boston: Allyn & Bacon, 1969.

Jarrett, Henry, ed. *Environmental Quality in a Growing Economy.* Baltimore: Johns Hopkins University Press, 1966.

Johnson, H. G., and Nobay, A. R., eds. *The Current Inflation.* New York: Macmillan Co., 1971.

Johnson, Warren A. and Hardesty, John A. *Economic Growth vs. the Environment.* Belmont, Calif.: Wadsworth Publishing Co., 1971.

Keynes, John Maynard. *Essays in Persuasion.* New York: Harcourt Brace, 1932.

Klein, Philip A. *The Management of Market-Oriented Economies.* Belmont, Calif.: Wadsworth Publishing Co., 1973.

Pechman, Joseph A. *Federal Tax Policy.* Rev. ed. Washington, D.C.: Brookings Institution, 1971.

Robinson, Joan. *Freedom and Necessity.* New York: Random House, Vintage Books, 1971.

Samuelson, Paul A. *Economics.* 6th and 9th editions. New York: McGraw-Hill, 1970 and 1973.

Schultze, Charles L.; Fried, Edward R.; Rivlin, Alice M.; and Teeters, Nancy H. *Setting National Priorities: The 1973 Budget.* Washington, D.C.; Brookings Institution, 1972.

Shultz, George P., and Aliber, Robert Z., eds. *Guidelines, Informal Controls, and the Market Place: Policy Choices in a Full Employment Economy.* Chicago: University of Chicago Press, 1966.

Smith, Adam. *The Wealth of Nations.* New York: Random House, Modern Library, 1937.

Stigler, George J. *The Theory of Price.* 3d ed. New York: Macmillan Co., 1966.

Teilhard de Chardin, Pierre. *The Phenomenon of Man.* New York: Harper & Row, Publishers, Harper Torchbooks, 1961.

Index

Behavior modification, 48–50, 128n6

Collective bargaining, 11, 20–22, 98–99
Collective demand, 60–61
Concentration of economic power: concentration ratio, 18; conglomerate movement, 14, 35–38; decrease in, 24; M. Friedman on market power, 84; in history of economic thought, 85; increase in, 84, 86; index of concentration, 127n*1*; interventionists on, 90–91; measures of, 86–89; noninterventionists on, 89–90; and price stability, 13; R. Solow on market power, 84
Conglomerate movement. *See* Concentration of economic power
Controls, social: of corporations, 4; of public utilities, 4
Controls: and American "incomes" policy, 28; M. Friedman on, 82–84; and Nixon administration, 28; of prices by producers, 13, 28; R. Solow on, 83–85; wage and price, 10, 11, 16, 22

Distribution of wealth and income: bottom fifth, 24–25; and buying power, 57; impact of PSP on, 25; impact on public goods, 61–62; orthodox economists on, 1–2; and political pragmatism, 56; redistribution vs. more growth, 58–59

Economies of scale, 131n*2*
Education: *Brown* vs. *Board of Education*, 2; fiscal benefits of, 47; costs and benefits of, 113–16. *See also* Pay-for-school plan
Emergency Employment Act of 1971, 29, 63, 118
Evolution: and economics, 51, 67–68; and leisure time, 68

Excess profits tax, 27–28, 35
Externalities. *See* Pollution

Federal Reserve: banks, 12; and goal of price stability, 54; system, 29
Freedom and control: J. M. Clark on, 92–93; M. Friedman on, 93; J. K. Galbraith on, 93, 94; F. A. Hayek on, 92; Philip A. Klein on, 95; Joan Robinson on, 95
Full employment: Act of 1946, 29, 56; and inflation, 1–5, 16; as a measure of level of performance, 54; relationship to welfare costs, 116–17

Government. *See* Role of government
Growth, economic: declining rates of, 71, 74–77; and inflation, 2–3; as a measure of level of performance, 54; and pollution, 69–71. *See also* Profit Stabilization Principle
Guaranteed income: and economic stability, 76; family assistance plan of Nixon, 111–12; various types, 109–17

Inflation: British control of, 103–5; A. F. Burns on, 95–100; cost-push, 4–5, 10, 20; cost-push and PSP, 40–41; demand-pull, 4; expectations, 4; experience of 1958–64, 83. *See also* Full employment
Inflation/unemployment remedies: antitrust, 15; automatic monetary policy, 15; direct controls, 15
Interventionist economics: W. Adams, 90–91; European, 26; J. K. Galbraith, 10; R. M. Nixon, 10

Labor force participation rates: in general, 62; of teen-agers, 9; of women, 9

Marginal propensity to consume. *See* Multiplier effect
Medical care, 63
Military expenditures, 131n*1*
Multiplier effect: and employment, 122–23; and marginal propensity to consume, 122–23, 130n*5*; and tax cuts, 129n*5*

Negative income tax, 41
Noninterventionists: M. Friedman, 11–12, 89–90; J. K. Galbraith, 90–91

Optimal population. *See* Population
Optimal production. *See* Production
Orthodox economics: American view, 1; assumption of insatiability, 58; assumption that work is onerous, 62; economic growth, 1, 3; excess profits tax, 28; investment bias, 59–60

Paretian test, 130n*7*
Pay-for-school plan, 113–16
Personal income limitation, 76
Phillips curve, 4–7
Pollution, 53–54, 70–71, 131n*2*, 133n*4*
Population: distribution of, 134n*4*; growth of, 69–75; its impact on resources, 124–25; as investment stimulant, 72–73; optimal, 69–70, 124–25; stationary, 71–75, 134n*4*
Price elasticities, 24
Price stability. *See* Economic tests of levels of performance; Full employment; Profit Stabilization Principle (PSP)
Production: optimal, 70, 133n*3*; zero growth rate, 71, 74
Profit rate: defined, 18; as guideline for price increases, 17–18; for hypothetical company, 18–20; range for smaller companies, 23–24; for U.S. firms, 18. *See also* Profit Stabilization Principle (PSP); Profits
Profits: and capital consumption allowances, 32; changing concept of, 105; congressional control of, 27; defined, 16; as guideline to price increases, 28; and political campaigns, 50. *See also* Excess profits tax; Profit rate; Profit Stabilization Principle

Profit Stabilization Principle (PSP): and allocation of resources, 33; and behavior modification, 48–50; British plan, 103–5; and collective bargaining, 35; and competitive industries, 25; and conglomerates, 35–38; and cost-push inflation, 40–41, 44; and declining rates of economic growth, 77; defined, 17; and economic growth, 31; and fiscal policy, 41–43; and flights of capital, 39, 77–78; and gradual price increases, 127n*2*; for a hypothetical company, 18–22; impact on productivity gains, 23–24; and land prices, 25–26; and monetary policy, 43–44; and planning, 45–46; and price elasticity, 37; and shortages, 100–102; and social responsibility of corporations, 107; and unemployment problems, 34–35; and unincorporated businesses, 35

Public interest, principle of: defined, 64–65; and freedom, control, and planning, 91–95; international application of, 78; and laissez-faire, 65; and property rights, 130n*8*

Quantity of money, 12, 82

Radical economics, 1
Role of government: in the economy, 16–17, 28–29; in job creation, 117–24; in maintaining prosperity, 97–98; and manipulation of prices and profits, 49; and maximization principle, 46–47; and public ownership, 60–61; and tax cuts to stimulate economy, 59, 129n*5*

Scarcity of resources: and economics, 58; and population, 124–25
Social costs: average, 133n*3*; marginal, 132
Social responsibility of corporations, 105–9
Spaceship Earth, 79, 124–25
Stationary state, 78–79

Steel industry, 13, 17, 18
Structuralists: Walter Adams, 13

Tests of levels of performance: cost efficiency, 53; conflicting goals, 55–57; engineering efficiency, 52–53; full employment, 54; price stability, 54; rate of growth, 54; Veblen's views on, 53

Underemployment, 8
Unemployment: and conglomerates, 14; "desirable" rates, 8; European rates, 8; fitting jobs to the unemployed, 63; and M. Friedman, 12; trade-off with inflation, 4–7; U.S. rates, 5, 8, 9; wartime rates, 8; in 1920s and 1930s, 55. *See also* Inflation/unemployment remedies

Wage and price controls. *See* Controls, wage and price
War: cessation of, 79; Korean War and U.S. economy, 27, 28; military expenditures, 67
Women's movements, 68–69
Working week: reduction of, 63, 76–77
Working poor, 8

9043